THE FOOD AND HEAT PRODUCING

SOLAR GREENHOUSE

DESIGN
CONSTRUCTION OPERATION

FISHER YANDA

Library of Congress Catalog Card No. 76-47993
ISBN 0-912528-12-5

Published by
John Muir Publications
P.O. Box 613
Santa Fe, New Mexico 87501

Distributed by
Bookpeople
2940 Seventh Street
Berkeley, California 94710

Typeset by Barbara Luboff
Printed in the United States of America

First Printing, November 1976
Second Printing, April 1977
Third Printing, August 1977

TABLE OF CONTENTS

INTRODUCTION		1
CHAPTER I	THE GREENHOUSE BIOSPHERE	3
CHAPTER II	THE DEPENDENCE CYCLE	7
CHAPTER III	THE DESIGN	9
	Principles	9
	The Site	13
	Exterior Design	17
	Interior Design	21
CHAPTER IV	CONSTRUCTION	25
	General Tips	25
	The Site	27
	The Foundation	28
	Massive Walls	34
	Frame Walls (Clear and Opaque)	37
	Roof Rafters	41
	Painting the Frame	42
	Choosing the Clear Glazing	43
	Clear Walls	44
	Roof Installation	45
	The Interior Plastic	47
	Insulating Frame Walls	48
	Interior Paneling	48
	Doors and Vents	48
	Exterior Paneling and Insulating	49
	Sealing the Greenhouse	49
	Tools Needed	50
	Materials List for 10' x 16' Attached Solar Greenhouse	51
CHAPTER V	PERFORMANCE AND IMPROVEMENTS	53
	The Basic Unit	53
	Increasing Collection	55
	Increasing Storage	57
	Decreasing Losses	58
	Active Systems	60
	Conventional Heating	63

Cooling the Greenhouse . . . 63

CHAPTER VI THE GREENHOUSE GARDEN . . . 65
Greenhouse Layout . . . 65
Soil . . . 66
Hydroponics vs. Soil . . . 68
Fertilizers . . . 69
Planting Layout . . . 70
The Vegetable Planting Cycle . . . 73
Flowers in the Greenhouse . . . 77
Maintenance . . . 79
Bugs in Your Greenhouse . . . 81
People and Plants . . . 86

CHAPTER VII THE STATE OF THE ART . . . 87
Individuals . . . 87
Research Institutions and Organizations . . . 118
Manufacturers . . . 130

APPENDIX A SUN MOVEMENT CHARTS . . . 139

APPENDIX B PLANTING CHART . . . 145

APPENDIX C ONGOING RESEARCH . . . 149

BIBLIOGRAPHY . . . 159

This book spans three years hard work with about sixty greenhouses built or examined along the way. The final result couldn't occur with the efforts of only two individuals. Here are some of the people who helped make it happen:

All the contributors to Chapter VII
The Solar Sustenance Project Volunteers
Keith Haggard
Peter Van Dresser
Joan Loitz
Frances Tyson
Ken and Barbara Luboff
Jene Lyon for permission to reprint sections and drawings from *An Attached Solar Greenhouse*
The entire Robert Bunker family

Also the following farsighted agencies which have supported our efforts:
The Four Corners Regional Commission
The Energy Resources Board of the State of New Mexico

For graphic contribution to the book:
Los Alamos Scientific Laboratories–Q-Dot Division
Tucson Environmental Research Laboratory–University of Arizona

For being there whenever we needed them:
The New Mexico Solar Energy Association

Finally, the two *most important* persons:
Lisa Gray Fisher who put all our drafts together on her desk and pulled out a book. As a reborn greenhouse farmer, she also wrote much of Chapter VI.
Susan Bunker Yanda...the only woman I know who can simultaneously lay adobes, feed and care for two infants, order materials, teach physicists how to nail lumber, plan a garden layout and calm a hyper husband.

In Chapters I through VI, the illustrations were done by Rick Fisher and photographs by Bill Yanda unless otherwise noted.

INTRODUCTION

First, a definition is in order, as there is some confusion created by the term "solar greenhouse." The confusion is understandable since, by definition, all greenhouses are in fact, solar. However, traditional greenhouse design has rarely been concerned with the most effective use of the sun's energy. Those described in this book are. In their design and operation, we have incorporated three basic elements:

1. The most efficient collection of solar energy.
2. The storage of solar energy.
3. The prevention of heat loss during and following collection periods.

By attention to those elements, we reap the following benefits:

1. Surplus thermal energy produced in winter which can be used immediately in an adjoining structure or stored for later use.
2. Independence from mechanical heating and cooling devices powered by fossil fuels.
3. Utilization of an optimum amount of insulation and thermal storage on an efficient, cost-effectiveness basis.

This book, the designs and the subsequent benefits involved all come from a basic concern with people's relationship to their environment and the forces within it. Working on the premise that one basic environmental problem is centered around *misuse* of energy, we realized that, while many people wish for alternative systems, the success of such systems is totally dependent on the individual's commitment to the system coupled with an understanding of what makes it work. That means you, and we want you to know exactly what's involved in building and maintaining your own solar unit.

In the following pages, we've shown methods which can be used to make an appreciable addition to the quality of your life through a closer involvement with your food chain, fresher (and cheaper) vegetables, a free source of partial heating for your house, a more realistic integration with the cycles of the sun, the seasons, the weather and the world, and independence from corporate energy and food games. Whether or not you actually build a greenhouse depends on many factors: space, economics, appropriateness to your location and determination, to name a few. But even if you don't build, reading this book can deepen and enlarge your understanding of your environment and your relationship with it.

This book grew out of the Solar Sustenance Project and that in turn grew out of a basic concern with the matters mentioned above. The Project was a modest ($15,000) attempt to see if a sensibly designed attached greenhouse(s) could lengthen the pitifully short growing season in the mountains of northern New Mexico. The Project also had the goal of finding out how much, if any, heat produced by the greenhouse could be used by the adjoining home.

When we began the Project, some engineers and architects insisted that our simple greenhouse wouldn't lengthen the growing season even a week. We were told by others that the 90 degree heat produced by the units was virtually useless. Fortunately, we didn't listen to them. To balance the negativism of the cynics, we did have the support of many people in the field: Keith Haggard and Peter Van Dresser of Santa Fe, T.A. Lawand of the Brace Institute in Quebec, and several of the people mentioned in Chapter VII.

Through the Solar Sustenance Project, we provided eleven solar greenhouses for low-income families scattered throughout the New Mexico hills. We attached them to the side of any structure we could tie-in with and that wouldn't blow away in a good wind. Some of the families looked on them and us as more of a curiosity than a functional addition to their homes. That was before. Now, we get universally positive reactions from owners, tourists and interested bystanders.

Our work on the project and on this book is founded on two principles: the first that food production should be a low-energy process. The process is begun by growing as much as you can at home, avoiding anything that requires more units of energy to produce than it contains. For that reason, highly-controlled, close tolerance food production techniques relying on outside energy sources to maintain them are not included in our work.

The second principle is that greenhouses and other habitable structures should be designed to make maximum use of natural energy flow and to make minimum use of fossil fuels. This means designing a "passive" structure with proper orientation, thick walls (high mass) and good insulation. This is not a new idea, but it is being re-examined today in the light of present technological capabilities. While a passive structure delivers obvious benefits, it also demands a great deal more thought, design work, labor and care in building.

In many ways the passively designed structure is in direct opposition to the current American mode of living. It's not temporary by nature. The temperatures fluctuate--it doesn't remain 72 degrees night and day: the structure itself has a "thermal momentum" that is much like the physiological processes of a human body, charging and discharging, inhaling and exhaling. Most importantly, a well-designed passive structure doesn't depend on a constant supply of energy to keep it liveable. The building uses the sun as the Earth does, only better.

Although the units we've presented are designed specifically for the dry, high-altitude, high-sunshine Rockies, the principles which make them work are valid anywhere in the world. Depending on where you live, you may need to increase the performance of your unit through modifications in design or addition of more sophisticated heat collection and storage systems. For those to whom this applies, we've presented a wide range of such improvements in Chapters V and VII.

If you decide to build and operate a solar greenhouse of your own, you will be joining a group of experimenters in what is still an infant science. You do not need to be a scientist to participate. All the principles involved are elementary and logical. Their simplicity makes the benefits derived from becoming an active member of the solar community easily accessible to you. Welcome.

CHAPTER I The Greenhouse Biosphere

The concepts of *environment* and *ecosystem* have been around for a long time, but only in the past few years have these ideas become part of the public awareness. Many of us only realized the profound implications of these concepts when we saw the first photographs of the earth taken from space by the astronauts. The earth is indeed a closed system, one that must sustain itself through a harmonious balance of its elements.

When you build your greenhouse, you will be creating a very special space, an earth in microcosm. You will control the character of the space to a great extent. Your imagination and design will determine how well the natural life force sustains itself and what you derive from it in return. You are, in effect, producing a *living* place that will grow and evolve with a life force of its own.

The special environment that you will create is called a *biosphere*. Webster's definition of a biosphere is: "A part of the world in which life can exist...living beings being together with their environment." As a living being, you are an essential element in maintaining your biosphere. Sowing seeds, nurturing the earth, watering, fertilizing plants and soil, and controlling the temperature and humidity will be your contribution to the biosphere. The greenhouse will reward you with the personal fulfillment of living within the cycle of growth.

FIGURE 1

Biospheres vary greatly in the number of their components and life systems, depending upon the interest, time and energy invested in them. A simple, easily maintained example would consist of a small structure with a few planting areas. Closely related, hardy varieties of vegetables and/or flowers would be chosen for cultivation. As their needs are similar, they would not require a great deal of time or attention. You may, however, prefer the role of maintaining a complex biosphere containing a wide variety of life forms. Some experimental units of this type combine plant growth (soil or nutri-culture) with the production of animal protein in the form of fish and rabbits. These systems attempt to achieve a symbiotic balance between the various organisms, using the by-products and waste of each to support the other. The more complex environments may also employ wind generators to power independent heat collectors, sophisticated storage facilities and other improvements (Chapter VII). These systems obviously demand much more time, attention and a strong interest in experimentation.

As a living space, your biosphere will grow and affect things around it. If it is attached to your house or another structure, an interaction between the two will occur. The conditions that develop in the greenhouse will be shared with an adjacent room or building in the forms of heat, humidity and the exhilarating fragrance of growth. In addition to pure sensuous delight, benefits can be realized in an economic sense as well as through a reduction in heating costs and food bills. The changing moods of the life system will soon become evident and you may find yourself reacting to them much as you would to a human personality.

FIGURE 2

Along with these rewards are the health benefits that you will enjoy. Greenhouse-fresh produce, especially if it is organically grown, can be far superior to its supermarket counterpart. Commercially produced foods may contain harmful chemicals, and in many cases lose much of their food value during the days they are in transit and on the shelf. Not only will your body welcome the added nutrition of home-grown produce, but you will also experience an unbelievable increase in flavor from the fresh vegetables. The environment of the greenhouse can also produce a feeling of well-being and tranquility. It may become a spiritual refuge from the outside world.

Perhaps the most dynamic aspect of your newly created biosphere is its relationship to the life force outside of our earth's environment—the sun. Solar energy affects every facet of life and change on earth. The sun produces movement in the atmosphere, water and land masses. It acts upon the earth's orbit and seasonal changes. Its waves of visible and invisible energy are the basis of all growth and life. This awesome force will be the medium through which you work. You will collect its energy, contain and store it, alter and direct it in the way most beneficial to the support of your biosphere. The sun will combine with air, earth and water to produce the fifth essential element in the greenhouse, plant life. In the management of your biosphere, your role will be to complete this five-sided cycle.

CHAPTER II The Dependence Cycle

The mass-market age is the mass-dependence age. Dangerous aspects of the dependence cycle are self-evident. Dependence is addiction. Whether it's a dock loader's strike in Philadelphia or a two cent jump in the per-gallon price of gasoline, the result is the same. Changes are made in your life, usually for the worse, without your having any say in the matter. Urbanization is part of this cycle; specialization in employment is as well. Everyone in this country has felt the effects of this situation and suffered some of the consequences. When those consequences affect basic life functions, it becomes a serious problem. The question is, "How do *you* break the dependence cycle?"

Going back to the land is one method, but for the majority of people, those who live and work in urban areas, this isn't a viable alternative. Rural life isn't everyone's dream and it's difficult, to say the least, to turn a 40' x 80' city lot into a self-sufficient farm. But one doesn't need to be *entirely* dependent on the system. A greenhouse makes it possible to grow a substantial amount of food in a very small area. Moreover, it lengthens the growing season tremendously in most parts of the country as well as protecting crops from damage by hail, wind and animals.

In order to prevent trading dependence on one part of the cycle for another, a basic rule of thumb is to make a careful evaluation of how much energy goes into food production from seed to table, then compare that with the amount of energy that comes out of the food to an animal or person. Think about how much energy it takes to grow, harvest, pack, store and ship the lettuce in your salad and you'll quickly see what that means. Consider gasoline and oil for tractors and trucks, energy expended to drill that oil, to transport roughnecks to the oil fields, to generate the electricity used in supermarket freezers and lighting, and on. And on. It adds up. Obviously a thoughtful long-range food/energy view takes production techniques into consideration, giving top priority to "low-energy-in, high-energy-out" approaches.

Again we come back to the family or community-operated greenhouse. It's hard to find a better example. It short cuts the entire process. The family that grows a head of lettuce realizes a measurable petro-chemical savings. Shipping costs are eliminated. Food is eaten fresh from the earth; no processing or packaging costs are involved. And it is produced by human labor without machine (purchase, operation and maintenance) expenses.

Aside from economic benefits, there is always the fact of quality food, fresh and healthy with amazingly good flavor. The pleasure of raising your own food ecologically and a feeling of self-reliance are additional rewards.

For all the above reasons, private greenhouse sales have increased tremendously. The problem with buying prefabricated greenhouses or plans is that they were designed without regard for the specific climate and solar conditions in your region, and they weren't planned for your site or your house. In fact, the majority of prefab greenhouses are designed as free-standing structures and demand additional fossil fuel in winter. Rather than adding heat to your home, they actually increase your consumption of fuel.

While we obviously haven't been able to see your home or your site, we've provided enough basics along with design modifications and information on how to use them, that you'll be able to use this book, save some money, understand why your greenhouse is working and best of all, end up with a life support system custom designed for your home.

CHAPTER III The Design

PRINCIPLES

The principles involved in the dynamics of a solar greenhouse are shared by all solar applications. Factors that apply specifically to an attached greenhouse will be discussed here.

Solar Radiation. In the context of a greenhouse, radiation or *insolation*, is energy that arrives from the sun in the form of short waves. Its intensity and duration are affected by a multitude of factors (latitude, smog, time of day, etc.).

Radiation. Heat energy is reradiated in the form of long waves from objects that sunlight strikes. These long waves heat objects and air molecules in the greenhouse and do not readily return through the glazed (glass and fiberglass) surfaces. This is known as the "greenhouse effect."

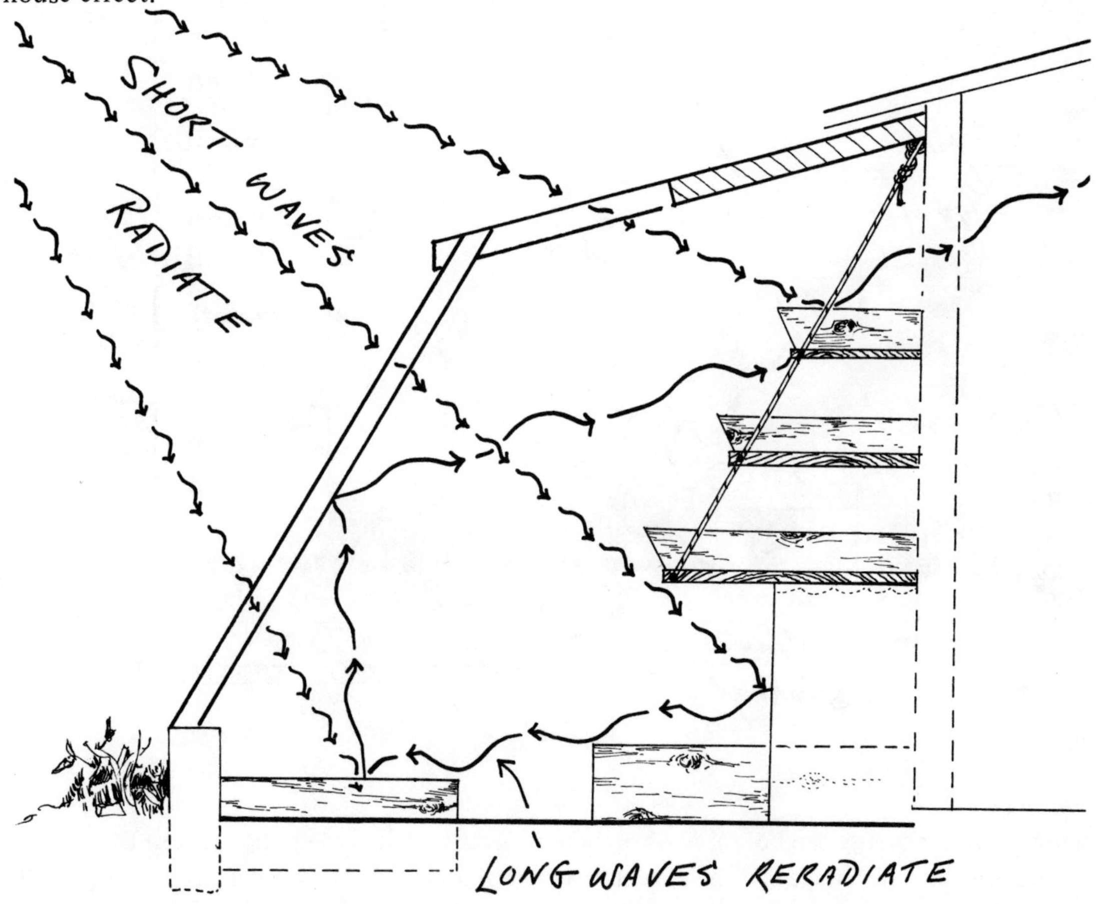

FIGURE 4

Collection. This applies to the gathering of radiant energy that falls on an object. The greenhouse may be considered a solar collector for the house. For the purposes of this book, the terms *collector surface* or *clear glazing* and *skin* will refer to the clear areas of the greenhouse that transmit (pass through) short wave radiation.

Angle of Incidence. This refers to the angle at which the sun's rays strike a collector surface. On any fixed surface this angle changes every moment of the year. Solar radiation is most effectively transmitted at an angle of incidence perpendicular (*normal*) to the collector surface. Besides the angle of incidence, the amount of energy transmitted depends on the physical properties of the skin (i.e., iron content in glass) and the number of layers in the glazing.

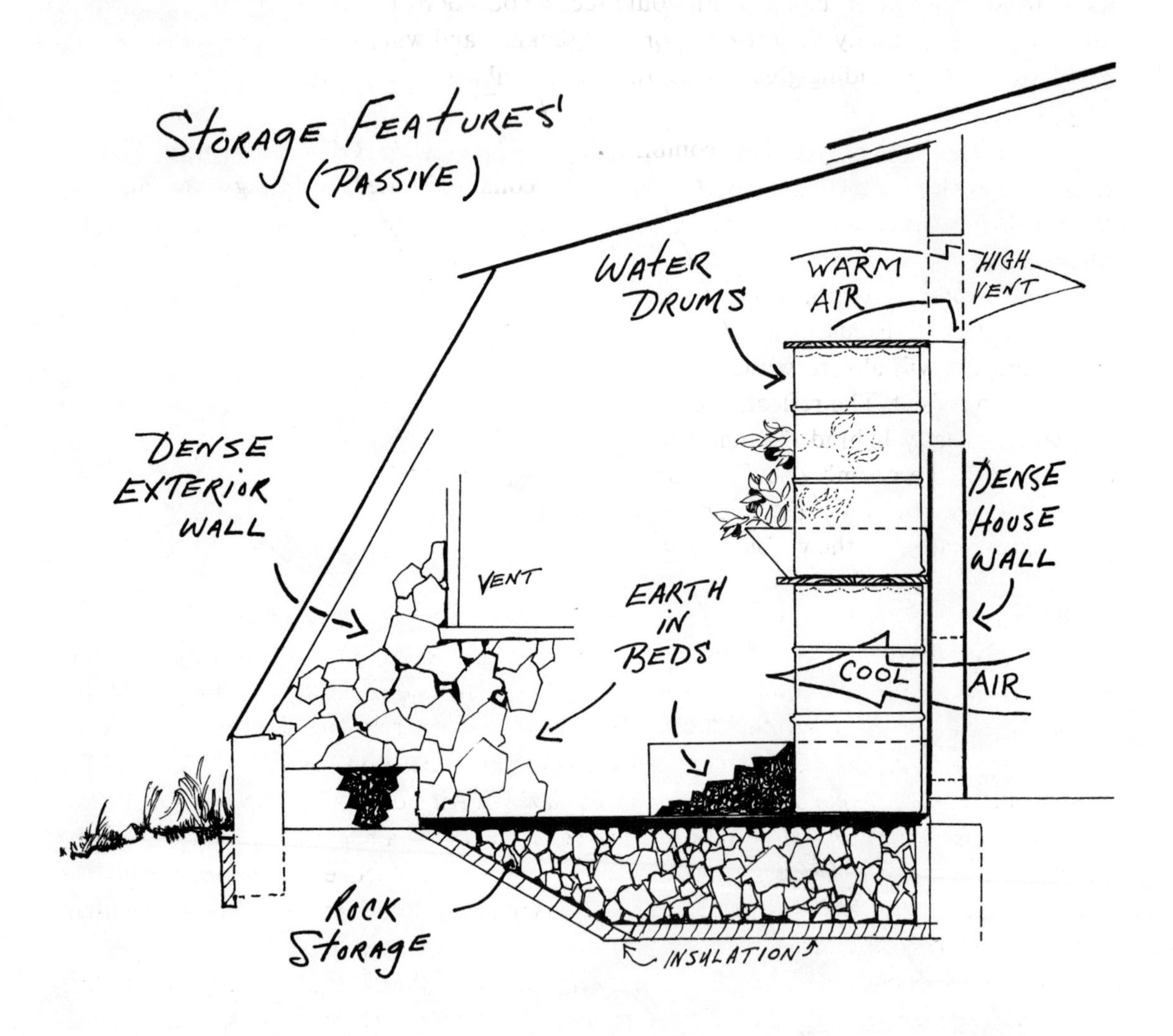

FIGURE 5

Storage. Solar energy can be stored directly in objects that it strikes. The amount of energy stored, the rate of storage and the rate of heat reradiated from the object depends upon the surface texture, color, thermal mass (density x specific heat), conductivity of the material and the temperature of the surrounding air. Common storage media used in a greenhouse are water, rocks and earth. We recommend two gallons of enclosed water or 80 pounds of masonry (rocks, bricks, etc.) per 1 square foot of clear glazing. For maximum effect, they should be visible to the winter sun. See Figure on the preceding page.

Absorption. Dark, low-reflective surfaces absorb more total radiation than light, reflective ones.

Reflection of light. As the north wall of an attached greenhouse is solid and does not transmit light, it is necessary to reflect (or bounce) some light from it in order to duplicate the naturally diffused light that a plant would receive outdoors. If this is not done, the plants can become abnormally *phototropic*, or light-seeking, and will not exhibit healthy growth patterns. In a freestanding greenhouse, the north wall can be tilted to reflect more light to the plants.

Dark and opaque surfaces combined with heat storage are required in the solar greenhouse to absorb and conserve heat, while light and clear surfaces are essential for healthy plant growth. The solution to these conflicting needs is a compromise. Some surfaces will absorb while others will insulate and reflect. The reflective areas can be placed directly behind the plants on the north side of the greenhouse. The chart in Figure 6 shows the reflective properties of various flat surfaces in the visible light range.

Visible Reflectance of commonly used reflecting materials

Material	Reflectance (Percent)
WHITE PLASTER	90 - 92%
MIRRORED GLASS	80 - 90%
MATTE WHITE PAINT	75 - 90%
PORCELAIN ENAMEL	60 - 90%
POLISHED ALUMINUM	60 - 70%
ALUMINUM PAINT	60 - 70%
STAINLESS STEEL	55 - 65%

FIGURE 6

Heat loss. Heat moves to cold, regardless of the direction it has to go. It is indifferent to "up" or "down", "inside" or "out." There are several types of heat loss to be concerned with in the greenhouse. One type is direct transfer, or **conduction** of heat from warmer regions to cooler ones. In the greenhouse, heat energy from warm inside air will be transferred through the skin to the cooler outside air. The process is slowed down by a "dead" air space between the glass or double glazings. Air pockets in fiberglas batt, styrofoam, sawdust or other types of insulation constitute dead air space and are the primary insulators in most insulating materials. They slow down conductive heat losses but do not stop them entirely. The ability of a material to insulate is rated as an "R" factor. The higher the "R" factor, the better the insulator.

Nocturnal or **clear sky radiation** is another source of heat loss. Warm earthly bodies lose their heat to the night (and day) sky...to space. Cloud cover, fog and haze are natural barriers to clear sky radiation losses. Smog is an unnatural, but effective, barrier. The loss occurs most rapidly through clear surfaces. This is particularly important in the Rockies and throughout the Southwest on crystal clear nights.

Convection is a third source of heat loss. An insulating barrier will lose its efficiency if there is air circulation within it. The faster the air movement, the greater the heat loss.

This circulation cannot be stopped entirely, but it can be cut down to preserve heat. You can think of **conduction** as heat moving through solids and **convection** as heat moving through fluids and gasses.

Infiltration losses are leaks around doors and vents, or through cracks and sloppy joints. They allow rapid heat loss and cold drafts and cannot be tolerated in a solar greenhouse. Air leaks are a major heat loss in buildings and the easiest ones to stop.

Evaporation could be a major source of heat loss but it should not be an important factor in a properly watered greenhouse. In the summer, evaporation will be used to help cool your greenhouse.

All heat losses are greater in the higher (warmer) areas of the greenhouse, through the clear walls and around the perimeter of the structure.

Air circulation. Heated air in the greenhouse rises and flows into a high opening to the home. A low opening in the shared wall allows cool air from the house to enter the greenhouse for heating. Without any mechanical devices this natural cycle will function continuously on any relatively sunny day (see this figure as well as figure 5 on page 10). Two additional benefits are provided by this type of air circulation. The plants in the unit convert carbon dioxide into oxygen-rich air for the home, a definite health benefit for the occupants. Also, in areas of the country with very low humidity, the added moisture from the greenhouse will be welcome in the home. The circulating pattern becomes better as the vertical distance between the high and low vents to the home is increased. Try to get a 6 foot vertical distance between the high and low vents. The same is true with summer venting. The total square footage of exterior vents should be about one sixth of the floor area of the greenhouse. The high vent is one third larger than the lower one. The exterior door counts as a vent. For example, in a 160 square foot greenhouse we have:

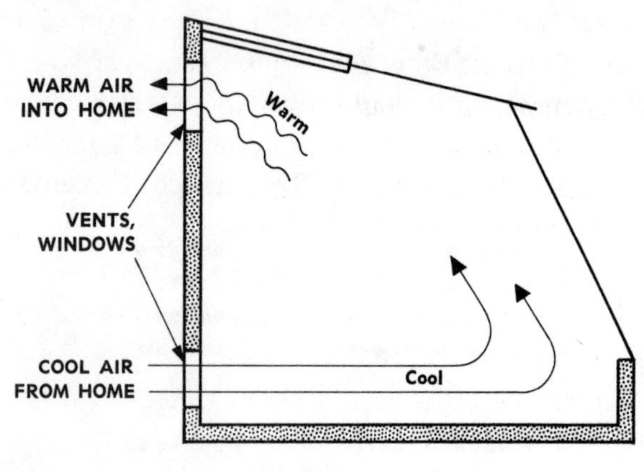

FIGURE 7

Exterior–	Low vent (16 inches off the ground) 2 x 2’	=	4 square feet
	High vent (7 to 8 feet off the ground) 2 x 3’	=	6 square feet
	Door 3 x 6’	=	18 square feet
	Venting area		28 square feet

Venting to the home doesn’t have to be this large. (It may be greater if there is a door/window combination.) Try to have at least 4 square feet of low openings and 6 square feet of high openings for a greenhouse this size. Scale up or down for your own application.

THE SITE

The meaning of the term *trade off* will become apparent when you begin to select a site for the greenhouse. All the conditions are not likely to be ideal, but it is important that positive factors are emphasized and detrimental ones kept to a minimum.

The first step in choosing a site is to determine how your home and property are aligned in relation to solar movement and other natural elements. Stand on the south side of your house. Where did the sun come up today? Where will it set? What will its rising and setting positions be on December 22nd and June 22nd (the solstices) relative to your south wall? What are the prevailing winter and summer winds?

Orientation. A solar greenhouse requires at least a partial southern exposure. Find magnetic south by using a compass. A survey map can tell you how many degrees east or west (declination) from your site true south is. Add or subtract these degrees to find true south. The magnetic pole roams around a bit, so try to get a fairly recent survey map. When you establish true south, determine how far from a perpendicular to south your house wall is. This *east-west* axis will be the north wall of the greenhouse, and it can be as much as forty degrees off true east-west without losing an appreciable amount of winter sunlight. If you find that you have no home wall that is oriented within forty degrees of the east-west axis, consider a corner location. If no acceptable location is found adjoining the house, then an independent structure must be planned. All of the natural considerations apply to independent as well as to attached solar greenhouses. Use the charts in Appendix A of this book as aids in visualizing orientation, sun movement and obstructions at your location. They should prove helpful. The following two sections will help you use them.

Sun movement. The sun is constantly changing its path through the sky, dropping low on the horizon in the winter and rising to an overhead position in the summer. A solar greenhouse differs from most solar systems in that it is not necessary to obtain the maximum intensity and duration of sunlight throughout the entire year. It should be designed and located so that it receives the greatest possible amount and intensity of light during the winter, when daylight hours

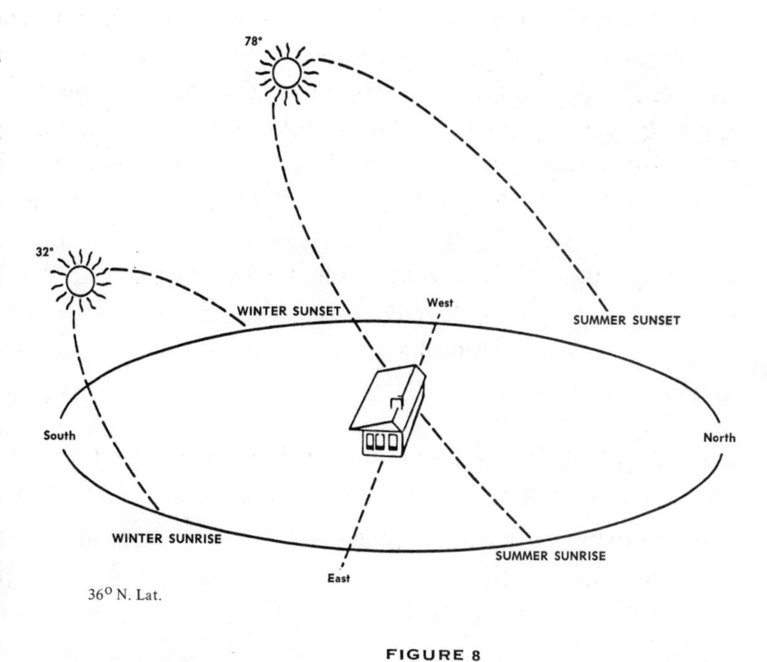

FIGURE 8

are few, and less light in the summer when overheating is a problem. The *photoperiod* (length of sunlight in a day) becomes particularly important for charging the thermal mass in the greenhouse during winter. Because the photoperiod is so short from October to March, both the plants and the heat storage features of the greenhouse need every available minute of it. Designing in accordance with sun movement patterns gives the solar greenhouse automatic advantages over conventional units for winter heating and summer cooling. This type of natural conservation has been widely neglected because of the availability of inexpensive heating fuels. However, pit and attached solar greenhouses were built in the late 1880's in New England and exhibited a precise understanding and use of sun movement.

The quality of light the greenhouse receives at various times of the year is also an important consideration. In Santa Fe, New Mexico, a winter morning is more likely to be clear than a winter afternoon. For this reason, it is advisable here to have a greater amount of clear wall on the eastern side of the greenhouse than on the west. Generally, an eastern clear wall makes sense in any locale. The greenhouse needs the early warming rays after a cold winter night. Joan Loitz (Chapter VI, page 81) claims that morning "is when the plants do their growin'. Give them eastern light."

Obstructions. Anything that blocks light from the greenhouse may be considered an obstruction. In examining an obstruction, its orientation and size are the two most important factors. Where is it in relation to the south face of the greenhouse? How much of the time is it blocking the sun and during what season? A deciduous (leaf shedding) tree that partially shades the greenhouse from the late afternoon summer sun could be an asset. A twenty-foot evergreen ten feet south of the unit is a serious problem. The charts in Appendix A, page 139, can be useful tools in determining how much sun will be blocked at various times of the year. For best midwinter operation, no more than an hour or so of midday sun can be lost to obstructions. In urban areas the possibility of a neighbor planting a tree or adding another story to his home directly in front of your greenhouse is something that should be considered. There is a great deal of legal research now underway pertaining to "sun rights." If in doubt, check into this possibility before you begin.

Your home may present a sun obstruction on the east or west side. While it might block some of the light, particularly in the summer months, it will also be blocking the wind and acting as an insulating barrier against the elements.

The important thing to remember is that an obstruction can be a positive or negative factor. You can compensate for some obstructions by proper bed layout. In other cases, exterior reflectors can make up, in intensity and duration, for light blocked by an obstruction. Remember that whenever light is sacrificed, the performance of the greenhouse is altered. Try to achieve the full winter photoperiod and at least ten hours of summer sun.

Wind. The natural flow of prevailing winds can be used to your advantage in the design of the greenhouse. In many parts of the country the summer-winter patterns will vary as much as ninety degrees. New Mexico has a pattern of winter winds from the northwest and summer breezes from the southwest. By mounting a low vent in the southwest corner of the greenhouse and a high vent in the northeast, the prevailing summer winds are used in natural cooling.

Another important matter is locating the *exterior* greenhouse entrance on the opposite side of the prevailing winter winds. In this way the entrance is partially protected from drastic heat losses when opened in the cold period. This is a necessity in a greenhouse that has no doorway to the home.

Drainage. Adequate drainage away from the home and the greenhouse structure is essential. Most buildings are constructed with a gradual slope away from them for runoff. When the extra roof area of the greenhouse is added, will this still be effective? Be sure to check the drains and gutters from the house and where they terminate. In urban locations it may be possible to connect to existing drainage lines. Try to have the ground runoff from the greenhouse follow the existing pattern of drainage. Some pick and shovel work may be necessary to facilitate this.

Percolation, the rate at which water can flow downward through the soil, is particularly important if you plan to have ground beds in the greenhouse. Although correct watering procedures would never allow saturation of the beds, accidents (such as leaving the hose on overnight) do happen. When soil conditions at the greenhouse site are not conducive to good percolation, you can add several inches of coarse sand or gravel to the ground level of the unit to aid drainage. Water accumulating under the floor is not beneficial to the plants or the thermal dynamics of the greenhouse. It lessens the effectiveness of any insulating barrier.

Utilities. It is convenient to have water and electricity available at the greenhouse site. A water faucet cuts down on the manual labor involved in hauling water to the plants, but as plants in a greenhouse do not require as much water as they would outdoors, their needs can be accommodated by hand. If a faucet is located at the site, plan to build it into the unit. In the spring and summer you can extend a hose through the door or vents for watering the outdoor plants. It is also possible to get an adaptor for indoor water outlets and run a hose to the greenhouse through a door or window.

An electrical outlet, like a water faucet, is convenient but not essential. It is enjoyable to have light for nighttime work, but difficult to justify the expense of the electrical power needed to light the structure in terms of the additional food it could produce.

Fans are very low power consumers and must be judged differently. In parts of the country such as the deep south, they may be a necessity for a successful year-round greenhouse. A fan would be a great aid in any area for moving warm air that the greenhouse produces in the winter into an adjoining home, thereby increasing the effect of the greenhouse as a collector. Here again, it might be possible to run an extension cord into the greenhouse and save the expense of permanently wiring the structure.

The main point is this: leave or design provisions for utilities if it is convenient and not costly to do so, but don't feel that you have to have them in order to have a successful solar greenhouse.

Openings to home. An attached greenhouse should have access to the house, ideally in several places. The reasons for this are both aesthetic and practical. If you find a south-facing wall that has a window and/or door that will be covered by the greenhouse, then you are very fortunate and have just saved yourself a lot of work. If not, don't despair; the situation can be remedied.

However, don't begin by punching holes in the wall of your home. This can be done at a much later date when you completely understand the air circulation patterns of the greenhouse and the heat that it is capable of producing. At this stage you are looking for locations that *do* or *could* have openings to the house.

A door from the home to the greenhouse allows easy access and integrates the two into a whole. A door from the kitchen to the greenhouse is particularly appealing and should be utilized if possible. Any windows between the structures are a delightful way of bringing the rich life forces of the greenhouse into the home throughout the year.

Keep in mind the visual characteristics of the building materials. If you plan to use fiberglass-acrylic panels or polyethylene in the greenhouse, remember that they are translucent, *not* transparent. Their visual appearance is roughly equivalent to that of a shower stall door. This might be an important consideration if you have a view you'd like to preserve.

Building codes. Building codes, inspectors and permits are strange inventions. Originally intended to be constructive, helpful devices, they can be restrictive, rigid and generally oppressive to innovative design work. The latest information in the code books about greenhouses was probably written around 1940. In some regions, greenhouses may be considered "temporary" structures (like gospel show tents) and have virtually no restrictions on their construction. In other areas, they may be subject to strict (and obsolete) codes.

The best advice is to find a friend involved in construction and check up on the "mood" of the codes and inspectors in your area. Quite possibly, the local inspector will be a considerate aid in your project, giving valuable advice on the strength of lumber, foundation footings, and so forth. If you are in doubt about the local situation, follow the prescribed code to the letter. In the long run this will be cheaper than tearing the structure down and doing it over.

EXTERIOR DESIGN

The exterior designs we show will be based on a lean-to or shed roof configuration, but the principles involved are applicable to any roof shape or design. A shed roof offers the following advantages:

1) Construction problems are reduced when working with standard lengths and angles. The fewer intersecting planes there are in a structure, the fewer junctions, the easier it is to seal. The addition can be tied into the existing structure fairly easily.

2) Both time and money are saved by the design. The spans, openings and panels can be planned around the United States building standard (four-by-eight-foot), so there is little waste. Material purchases are easy to estimate.

3) As major exterior planes are limited to four or five large areas, heat gain/loss calculations are a simple, straightforward process. When moveable insulation is utilized, the procedure of applying it is uncomplicated because you are working with large, flat surfaces.

In choosing a site, try to achieve the following:

1) Have as much south collecting area as is economically feasible.

2) Cover a long linear area of the house wall for storage and insulation. A long rectangular (east to west) greenhouse gains a greater photoperiod than a "boxy" or square design.

A rule of thumb that has proven successful in building solar greenhouses is to allow the length to be about 1½ - 2 times the width. A size that has been fairly standard in the Solar Sustenance units is 16' long by 10' wide (foundation extremes). A greenhouse of these dimensions has plenty of growing space with some room left over for working and relaxation areas.

If the width of the greenhouse becomes much greater than 10 feet and the pitch of the roof is shallow, rafters heavier than 2 x 4's must be used and the expense of building increases. A narrow greenhouse has other advantages in construction and in its thermal characteristics.

In the greenhouse shown in the diagram (Figure 9, following page), the sun at noon will strike no higher than point "A" throughout the winter. You can plan to add any direct-gain storage below and south of it. A wider greenhouse ("E") would mean that the clear area on the roof ("C") would have to be increased, creating more clear surface for heat loss in the winter and heat gain in the summer. Note that on June 22nd the area north of point "B" will be in the shade during the hottest part of the day. This means that your thermal

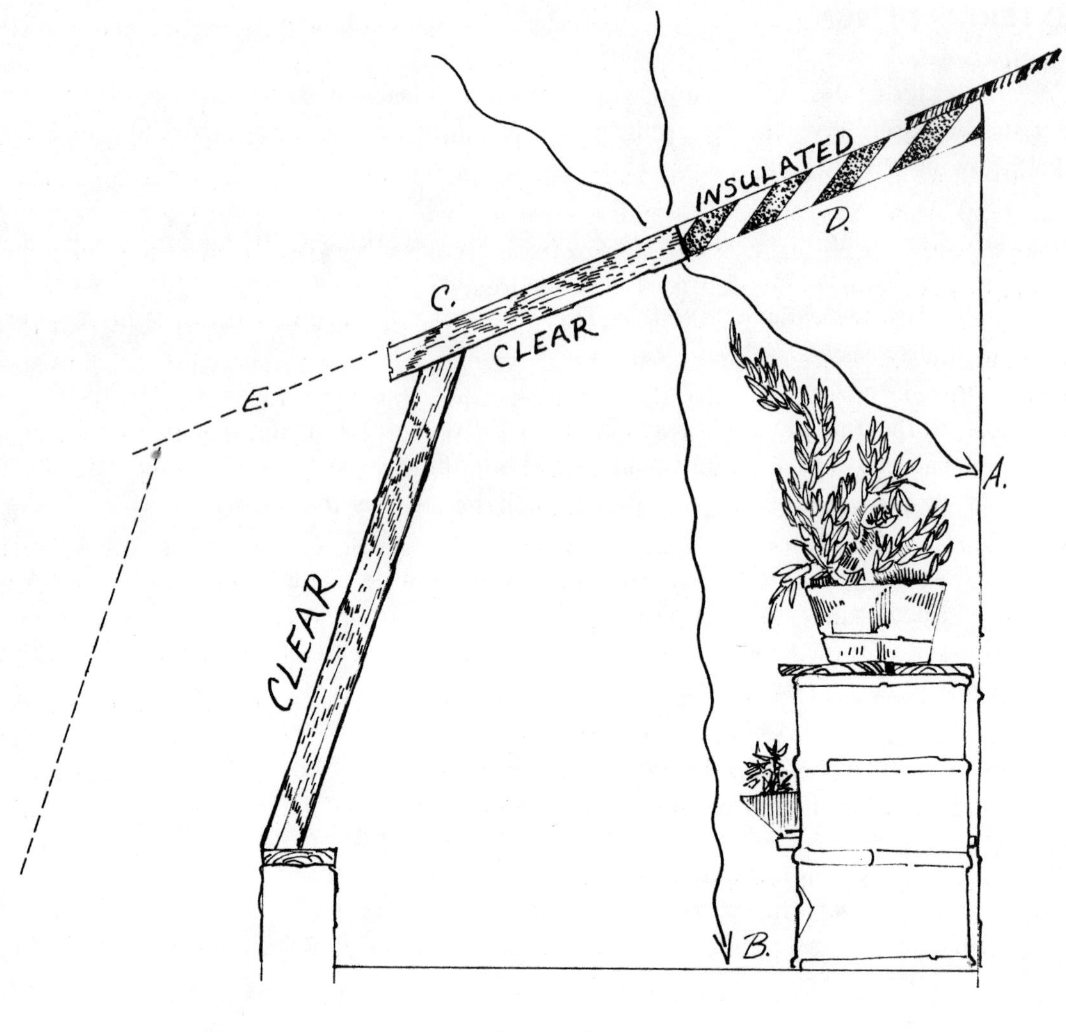

FIGURE 9

storage is out of the direct sunlight in mid-afternoon on summer days, when you don't need or want it heated.

The solid roof area of the greenhouse ("D") will be permanently insulated. The result will be a more even temperature range in the unit throughout the year. A partial sacrifice with built-in shading is that full light-loving plants (tomatoes, cucumbers, peppers) won't do as well in the rear of the greenhouse during the summer period. You can compensate for this by planting them in front. However, most flowers, shade-loving plants, and cool-weather vegetables will find the covered area of the greenhouse acceptable. The loss of a little summer light is more than returned in overall thermal performance in a correctly shaded greenhouse.

Plan your design to allow the winter sun to strike well up on the wall ("A") of the greenhouse (five to seven feet). This will guarantee that approximately one-third of the rear section will have partial shading in the heat of summer. The point on the rafters where the insulated roof stops and the clear roof begins can be determined by a cross section scale drawing using the information on the charts. (Appendix A, page 139).

Angle of the south face. In the solar greenhouse, you are working for maximum efficiency of collection during the winter period and reduced summer transmittance. To do that means that the angle of the south face should be close to a perpendicular, called *normal*, with the average "solar noon" angle of the sun during the coldest months. In the northern hemisphere this period is mid-November to mid-February.

If we average the winter solar noon angles in the contiguous U.S. we find that the optimum tilt for winter collection begins at about 50° in the southern United States and rises to 70° around the Canadian border. Remember, this is at solar noon; all other times of the day the sun is at a lower altitude in the sky. The lower altitudes mean that the collector tilt can be raised for better transmittance through the entire day. A formula I've used for establishing the tilt of the south face is the latitude + 35°. You can see that if you live north of 45° latitude, the angle becomes very close to vertical. To have exactly the correct angle is not critical. The tilt of the glazing can be as much as 50° off of normal and still not lose an appreciable percentage of light transmittance.

What is perhaps more important, is that a steeply tilted south face is *not normal* to the summer sun. In other words, you don't have a huge clear area perpendicular to the intense summer radiation. The relatively small clear roof area of the shed designs (Figure 10) insures that you have less overheating problems in the warm months. The large south face at a steep tilt is far enough off normal to reflect a substantial amount of radiation in the summer.

There are other important considerations to determining the angle of the south face. If you are planning moveable insulation (rigid or curtains) on the clear surfaces, a vertical plane is definitely easier to cover than one tilted at sixty-five degrees. Another factor is that a vertical south wall provides more interior space than a tilted one (Figure 10).

On the other hand, a vertical south wall demands more materials and greater spans. Also, given a height limitation on the home, the tilted face has a greater surface area to winter normal than the vertical one. Both designs have a relatively small *clear* surface on the roof that is normal to the summer sun. In our units, we have used both configurations successfully.

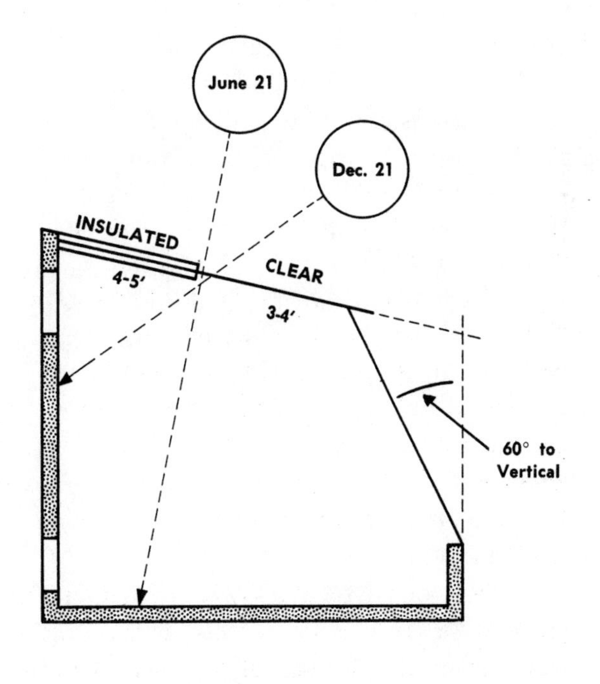

FIGURE 10

You may have an existing patio wall or fence that could serve as the south border of the greenhouse. The angle between it and the proposed apex might not be exactly correct for the maximum use of winter sun, but its availability can be a definite advantage. Make use of it if you can.

Insulating and storing walls. The solid walls in your greenhouse can be used for insulation, storage or best of all, both. Depending on your orientation, they will be built on the east or west side(s) of the unit and perhaps a low one will be included beneath the south collecting face. A typical *insulated* wall would be made of two-by-fours, with interior fiberglas or rockwool insulation, then sheathed, paneled and sealed. The advantages of such walls are that they are easy to build and inexpensive. This is the way the great majority of American homes have been built for the last four decades. The drawback is that if the heat is turned off for a time during the winter, the frame house gets cold very quickly. The structure is entirely dependent upon continuous heating.

Building a wall with thermal mass makes sense in any structure that uses direct sunlight for heat. The heat is stored in the building material and returned to the structure several hours later. Materials such as the adobe bricks used in the Southwest have the remarkable quality of delivering maximum stored heat about twelve hours after the peak collection period, when it is needed most. The old rock homes with thick walls found throughout much of the United States perform the same function. This natural cycle also works to the benefit of the occupants in summer, helping to keep the home cool in the day and warm at night.

Common massive building materials are ceramic brick, stone, adobe (fired or unfired), poured concrete and pumice (cinder) blocks filled with concrete. These kinds of walls require a heavy foundation and take a little longer to build, but they are worth the effort in a solar greenhouse. The thicker the walls, the greater the thermal mass.

For massive walls to perform properly they must be insulated on the *exterior* surface. (See Chapter IV, page 49, for techniques). When this is done, the walls become in effect a structural "Thermos bottle," radiating most of the heat gained during the day back into the greenhouse at night.

It is important to remember that the greenhouse does not need to have perfect characteristics in all of the discussed categories to be successful. However, if one factor is lacking, it would be wise to compensate for this somewhere else. For instance, if you decide to build simple stud walls, it is advisable to increase the amount of thermal storage in some other area of the greenhouse.

INTERIOR DESIGN

The design of the greenhouse interior will depend in large part on your personal attitude toward its use. Many people enjoy a greenhouse that provides space for activities other than gardening. If this is your feeling, allow plenty of room for sitting and moving about. You may arrange the planting areas around a central living space or separate the two completely.

Most greenhouse owners prefer to make maximum use of interior space for growing plants. This is a more difficult design problem and demands consideration of several important factors.

Access. If your biosphere is built against a wall having an existing doorway, the door should open away from the greenhouse area. All exterior doors are built to open out. This will

FIGURE 11

allow you more freedom in arranging the interior space. It is likewise preferable to build vents that open to the outside or that slide.

Provide sufficient walking space in your floor plan for unrestricted access to all planting areas. Plants will tend to overhang the beds, so allow for growing room. At times your greenhouse may have to accommodate several people; one expanded area of a walkway will furnish the needed capacity (see Figure 12, following page).

Planting areas. Permanent beds may be dug directly into the greenhouse floor. If additional depth is desired, supporting sides built above ground level will hold more soil. It is important to estimate shadows that will be cast from plant-filled beds. Beds located at the rear of the greenhouse (away from the sun) may be built above ground level to prevent their being shaded by front plantings. In the small greenhouse, optimum use of vertical space is essential. This may be accomplished by adding shelves, hanging beds and planters.

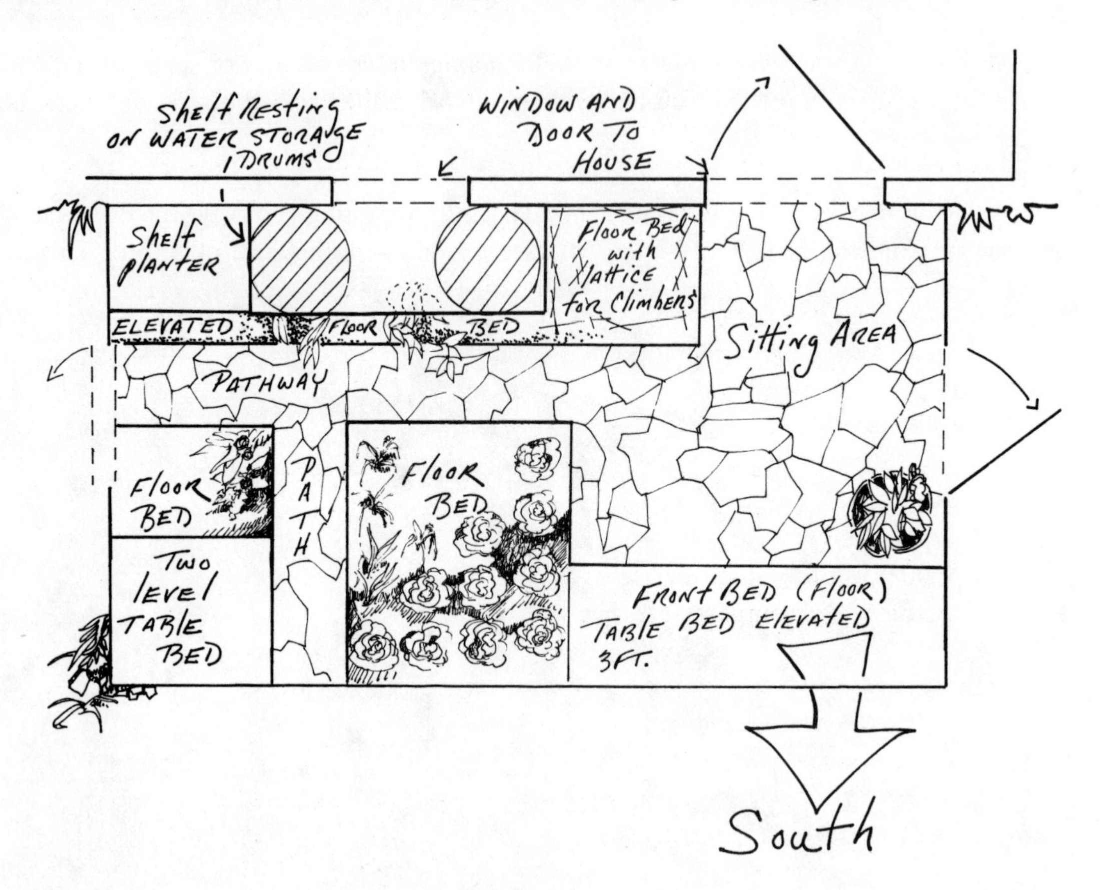

FIGURE 12

Table beds used in conjunction with ground beds will yield an even larger planting area. However, care should again be taken to insure that the front planters and tables do not unintentionally block direct sunlight to those in the rear. The above drawing illustrates a planting layout incorporating several of these features.

Including a water drum storage system into your greenhouse will require added design considerations from the outset. In terms of the interior design, your primary concern is the proper placement of the system. Exposed water drums, for instance, should be located such that they receive maximum direct sunlight yet do not shade your plants. Using the storage drums as shelf supports of planter bases is a good way to combine functions. If you plan to add this passive storage system later, allow adequate space in the original design to accommodate it.

Aesthetics. The main goal in designing your greenhouse interior is to produce an environment that is both practical for growing plants and beautiful to experience (for you and the plants). Your choice of an interior layout will to an extent be predetermined by functional considerations. Many designers equate function with beauty—the most functional is the most beautiful. Since the function and process of the greenhouse is such a positive, creative one, this notion will generally hold true. As you concentrate on producing the most functional space for growth, certain characteristics that are unique to the biosphere environment will guide you in deciding how the interior should look.

You may refer to Chapter VI for planting information to determine the most suitable growing arrangement for your purposes. Certain depths of soil and conditions of light and shade are recommended for different varieties of plants. These conditions will influence the dimensions and placement of beds and interior structures.

Growing, changing plants will be a primary part of your greenhouse. You may choose to heighten the profusion by using a wide variety of colors, shapes and surfaces in the walls and bed structures. On the other hand, you may wish to balance the active look of plant life with a more simple interior design by limiting the number of exposed (visible) materials to a few that appeal to you most. Choosing fewer colors will accomplish the same end.

If a constantly changing environment suits you best, build planting beds for growing "living walls" within the space. Climbing vegetables such as tomatoes or pole beans will cover a hanging string or stake lattice in short order. The planting will necessarily demand considerations of shading, but the aesthetic rewards can be great.

FIGURE 13

The finishing touches you apply are a matter of personal taste. Surfaces which receive direct sunlight and have sufficient thermal mass for heat storage should be painted with a dark nonreflective color. Flat black paint is most effective for absorption, but we also recommend a dark blue, brown, red or green. All insulating walls having no thermal storage potential should be a light color to reflect more light to the plants. Trim, such as wood stripping, molding or lath can be used attractively to unify planes (and to cover less-than-tight joins).

Perhaps the single most important step in designing the greenhouse interior is to examine your own likes and dislikes. The qualities of environment that you feel are most attractive should be combined in the plan. The value of this point has been proven in the Solar Sustenance greenhouses: if you are pleased with the final results, you will spend many pleasant hours in the environment, helping to insure its success and deriving the fullest benefits from the fruits of your labor.

FIGURE 14

CHAPTER IV Construction

In this chapter we will take you from the first shovel of excavated earth to the last nail in the wall. Experienced builders may want to skim this material for pertinent greenhouse information and do the rest of the building their own way. A list of tools and materials used in building a 10 x 16 foot attached solar greenhouse is shown at the end of this chapter. But read the entire chapter before you buy or build anything.

If you've never really built anything before, you're in for some surprises. First, there is nothing mystical about construction—houses, chicken coops, greenhouses, airports; the principles are all the same. Your potential building skill is as great as any builder, better than most, because it's *your* project and you care about it.

The main difference between you and building contractors is that they know how to cover mistakes. Almost any error can be corrected (or hidden). True, you must stick to some fundamentals to have a sound structure, but common sense rather than supernatural powers guides you in following these basic procedures.

There are about a thousand ways to do any particular building operation (that's what perpetuates the mystical aura that surrounds construction). Any "expert" will give you one or two of these ways. So will we.

Once you start building, you'll find it addictive. We have had a love-hate relationship with it for years. It will torture your days and keep you awake nights, but you'll want to do more. You will think of all the improvements that you could make in your house and you'll be hooked.

GENERAL TIPS

1. Plan each step of construction as completely as possible. Determine and obtain materials for each step before beginning. This is easy to say and not so easy to do. But if you consider the amount of time, organization and gas money that goes into a "quick trip to the supply store," you should be convinced of the need to plan ahead.

2. Many construction materials come in standard sizes. Sheetrock and plywood, for instance, come in 4' x 8' panels. Designing the dimensions of the greenhouse to correspond to these standard sizes can eliminate time-consuming cutting and custom fitting. It also reduces expensive waste.

   Do not expect framing lumber to be the size it's called. Those days are long gone. For instance: a 2x4 is 1½ x 3½", a 1x6 is ¾" x 5½". Also, when buying stud lumber, check the length. Not long ago, I bought some 8' studs that shrunk to 7'-7½" by the time I got home. When I called the company, I was told that including the top and bottom plates, the studs would make an 8 foot wall. Well, that's true, but I paid for eight feet!

3. Prices vary greatly. In shopping for materials, a few phone calls to competitive suppliers may provide substantial savings. If you can get some other folks interested in building a greenhouse, you can save money by making "quantity orders."

Americans don't realize that there is *no set price* on anything. Don't be afraid to bargain-shop or haggle. Ask for a special price. Get to know the manager of the store. Tell that person that you're working on an "experimental project." That always sounds interesting. It may bring the store increased business if your friends like your greenhouse and decide to build one themselves. Supply and demand still functions in the building industry. If an item isn't moving, the manager can lower the price on the spot. I once got a $350 table saw plus all optional attachments for $200.

4. You may choose to use less than "first grade" materials. Resawn or rough wood can be purchased for about one-third the cost of finished grade lumber. Useable materials are often discarded; you can recycle them. Check large construction sites, salvage yards and dumps. A note here: How much recycled material you use may depend on how much time you have to devote to this project. The reason most construction companies buy everything new and in standard sizes is to save time. If you have the time, save cash by scrounging.

5. In buying materials, order somewhat more than you expect to use. This will allow for mistakes and save unnecessary trips to the supplier. It's also wise to expect the total cost of construction to be somewhat higher than your estimate and the time involved longer. I'm usually off by about twenty percent. This is probably why the building industry has the highest percentage of new company failures of any industry in the country. When was the last time that you heard of a construction project being completed in less time and at a lower cost than the estimate?

6. Set up a staging and storage area for the materials. Plan to keep the entire building area off-limits to little children and pets. There are so many activities going on at a construction site; it's easy for someone to get hurt.

7. In all steps of construction, measure as accurately as possible. If in doubt, exceed the correct measurement rather than cutting under it. You can always take a bit more off but it's difficult to put it back on.

8. Perhaps the most useful piece of advice for the novice builder is to ask for advice. The elderly, experienced salesperson at the local hardware store may be a fund of building knowledge. Don't hesitate to tap this valuable source.

THE SITE

The site that you have chosen for the greenhouse may demand attention before you can begin foundation work. In certain cases, a site that is not level can work to your advantage. If the terrain slopes away from the existing structure, for instance, you might consider "sinking" the floor level of the greenhouse (Figure 15). A good depth for attached solar

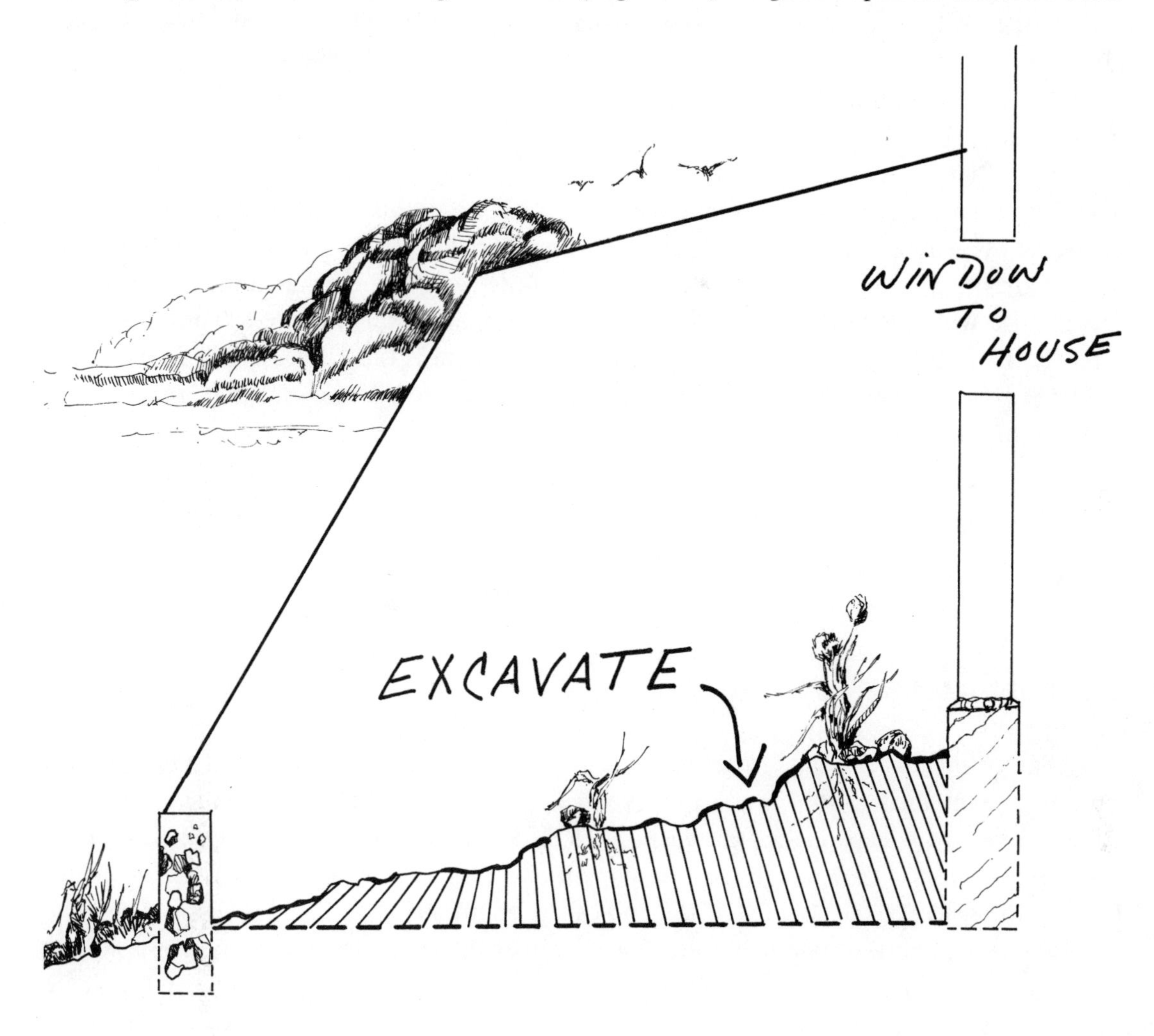

FIGURE 15

greenhouses is the same depth as the home foundation. One of the main reasons for excavating is to lower the "profile" of the unit so that it fits beneath the existing eaves of the house. Sinking the greenhouse will require more excavation than simply leveling the site, but it can result in more useable vertical space. Situating the highest point of the greenhouse interior directly adjacent to a house window or doorway will also supply more useable heat to the home.

If the ground slopes laterally to the side of the house, you may wish to design a split-level floor plan rather than level the entire site. Whether split-level, sunken or used in its existing state, the site should be relatively level (side to side, front to back) before beginning foundation work.

The excavation depth for a greenhouse is determined by several factors. Many people have the misconception that if you dig a little way into the earth, the below-grade soil will be thermal storage. Actually, one must go down a considerable distance below the frost line to reach earth that would constitute a heat gain for a winter greenhouse.

"Pit greenhouses" or "grow-holes" are based on this principle. They are dug out several feet below the frost line to enjoy the benefits of the earth's thermal storage. We have observed that grow-holes perform slightly better than solar greenhouses *only* in extremely cold weather (below -20°F in the New Mexico region).

To achieve increased performance in a pit type of greenhouse adjacent to the home, you might have to dig down several feet below the foundation of the dwelling. This is *not* advised. If you have an existing deep cellar or basement with strong walls and good drainage away from it, an attached grow-hole might do quite well. The hot air in the apex of the greenhouse would enter low in the home and the cool air in the basement would be circulated into the lower part of the greenhouse. The problem is that any design of this nature would require extensive excavation, landscaping and a thorough knowledge of the strength and condition of existing walls. It is not a recommended project for novice builders.

THE FOUNDATION

The foundation of any structure is one of its most important elements. If it is built properly, many future problems will be avoided. Careful measurements for the foundation are essential. We will assume for the purpose of these construction steps that you are building a greenhouse with a rectangular floor plan and that you are attaching it to the home. For the tools needed, see the list at the end of the chapter.

To determine the ninety degree corners off of the structure, place one edge of your framing square against the existing wall and extend the other edge with a string (see Figure 16 on following page). Stake the string at the distance you have determined for the outer boundary of the greenhouse. After repeating this procedure for the other end wall, measure to see that the two strings are parallel ("A" to "B" equals "C" to "D"). Connecting the outer perimeter stakes should produce a rectangle ("A" to "C" equals "B" to "D"). To double check your ninety degree corner angles, see that the diagonal measurements are equal ("A" to "D" equals "B" to "C").

We will describe the poured concrete/rock type of foundation because it is widely used and easily understood and constructed by the home builder.

Along the perimeter of the greenhouse excavate a trench to the desired width and depth. Make it at least 4 inches wider than the walls of the greenhouse and at least

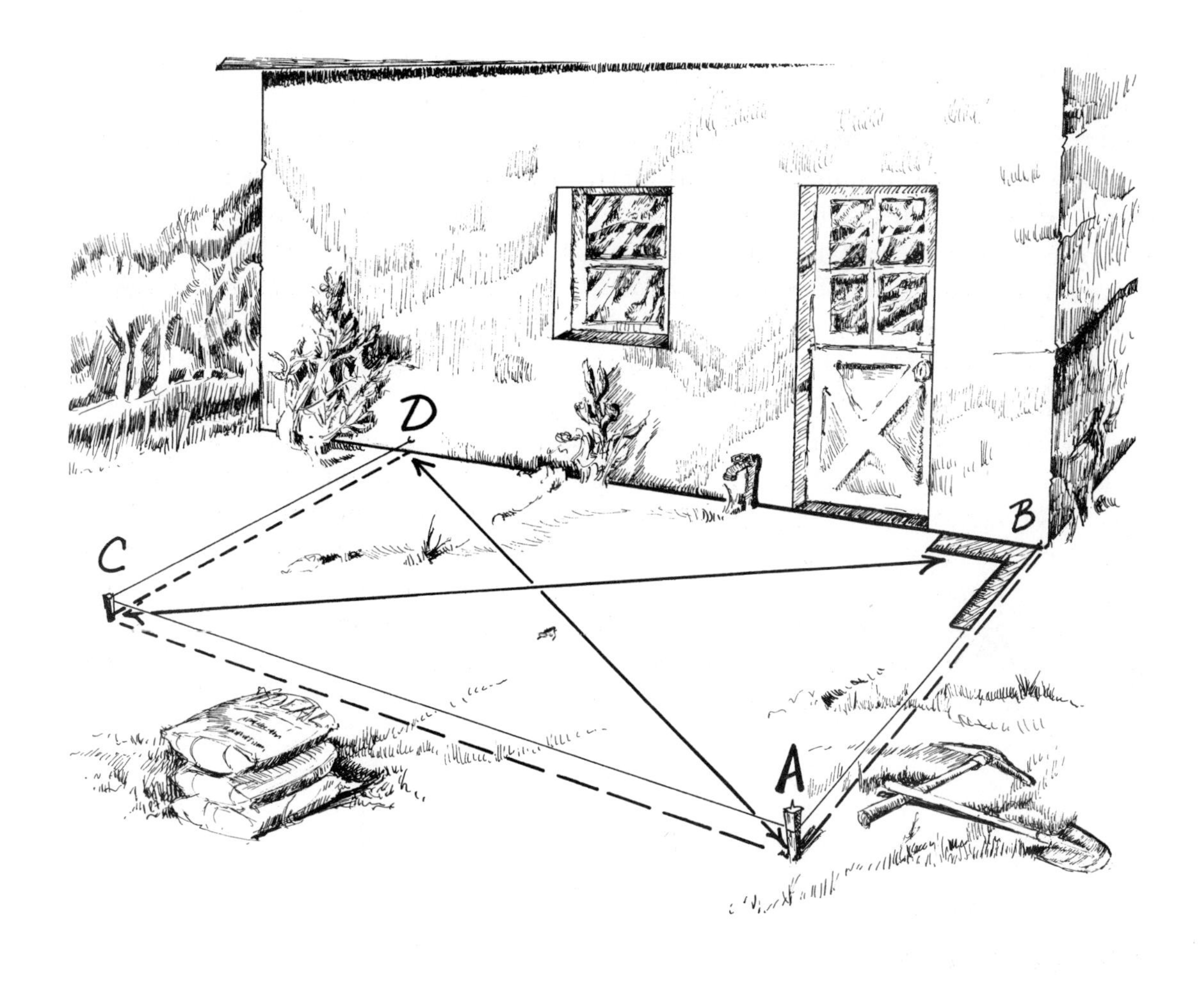

FIGURE 16

16 inches deep (if a massive wall(s) is planned). Drive stakes into the trench at 6 to 8 foot intervals, leaving 6 inches of the stakes exposed above the bottom of the trench. Check that the trench is level by laying a flat board from the top of the highest stake and taking a reading with the level from there. Do this around the perimeter of the trench. Fill when necessary; then smooth out the sides and bottom with a flat-nosed shovel. To double check the level, we recommend:

The Old Carpenter's Water Trick

> So, you want to check one end of the foundation trench with the other. You don't have a transit and the 2x4 won't bend. Get a friend. Then take a regular garden hose and lay it in the trench. Drive stakes in the end corners; each must be exactly the same height from the bottom of the trench. Holding the ends of the hose flush with the top of the stakes, fill it with water. If the extremes are level, the water level at each end will be equal.

The beauty of this trick is that it will work for any length and over rough terrain (with a couple of people and plenty of garden hose). Of course, the hose ends have to be held higher than any point in between.

After the trench is dug, leveled and cleaned out, keep all interested gawkers away from the edges so they don't cave in the sides.

Before the foundation is poured, the outside and bottom of the trench should be insulated with 1" or more of rigid styrofoam (see Figure below). Cut the panels to size and

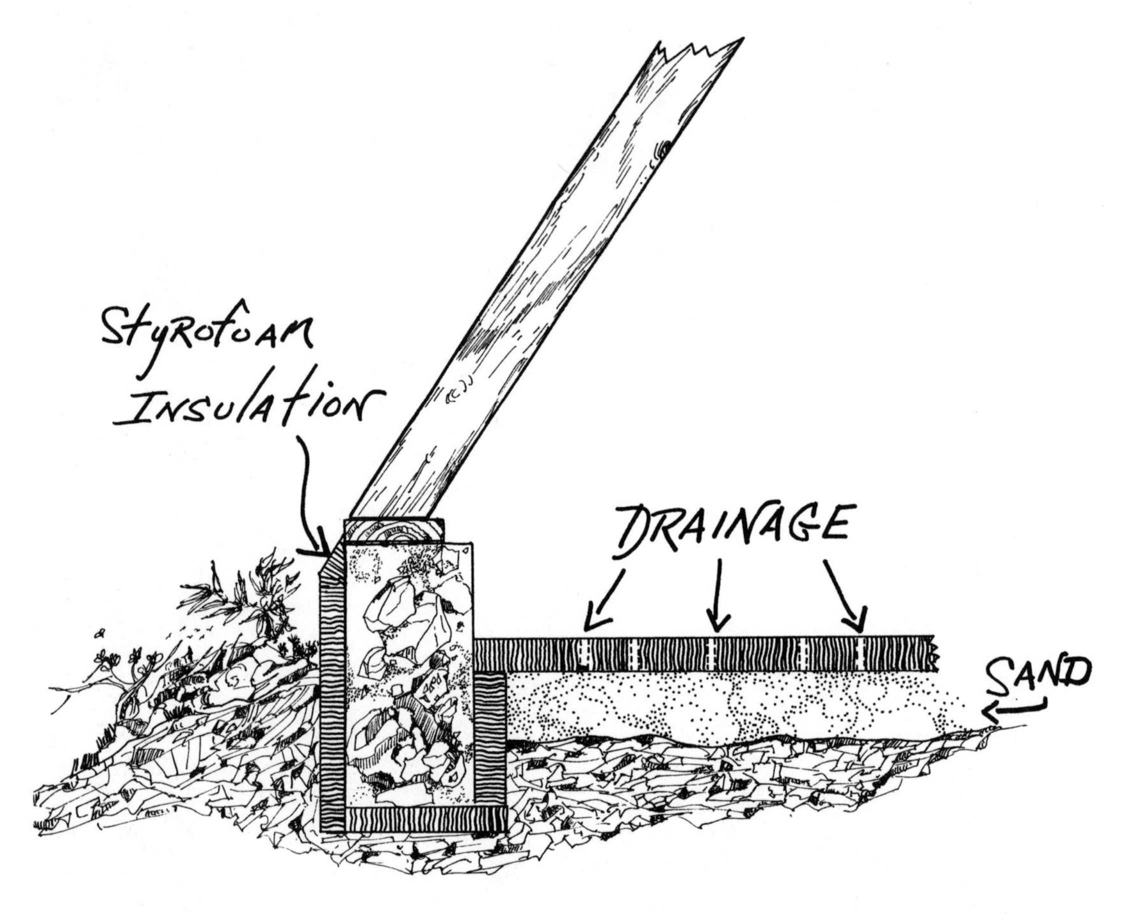

FIGURE 17

fit them into the trench. They can be temporarily propped in place until the concrete is poured. Another method of insulating the perimeter is to wait until the foundation has been poured and the concrete has hardened; then dig a trench around the outside of it. Line the trench with sheet plastic and fill it with sawdust, dry pumice or styrofoam beads. Enclose the loose insulating material with the plastic to keep it waterproof and cover the trench with dirt.

Another prepouring step is to insert reinforcing material in the foundation trench. If you have to meet stringent building code requirements, this may be mandatory. "Re-bar" or "re-rod," as it's called, can be used in 1/2" or 3/8" diameter. It can be bought and cut to length at any building supply store. Two lengths of re-bar are laid along the bottom of the trench about 8 to 10 inches apart, supported 4 to 5 inches off the bottom by rocks. Ends that meet are lashed together with baling wire.

I've poured foundations with and without re-bar. I often throw as many river rocks as I can find (5 to 8 inches in diameter) into the bottom of the trench and forget about the re-bar. I haven't noticed any settling or cracking in the foundations I've built this way.

The re-bar or no re-bar question reminds me of a typical bureaucratic hassle over the recent building of adobe homes in New Mexico pueblos. When the government engineers finally approved adobe for Indian housing (the all-adobe Taos pueblo has only been standing for a millennium or so), they stipulated that re-bar be inserted in the vertical walls every several feet. A Santa Clara Pueblo friend of mine said, "Isn't that going to frustrate the archeologists a thousand years from now? They'll wonder what all the little red holes are doing in the middle of those mud walls." It's probably true that the adobe walls will be standing when the steel re-bar has rusted out. Suit yourself about the use of reinforcing bar.

It's a good idea to bring the foundation above grade (3 or 4 inches). This automatically eliminates some drainage problems and is definitely necessary if frame, adobe or other water-soluble materials are going to be used to build the walls.

FIGURE 18 PHOTO BY KRISTEN MACKENZIE

Old lumber can be used for the forms to restrain any concrete that is above grade. Most anything will do to secure them in place; large rocks, blocks, stakes, wire. Be sure the forms are the right distance apart and *well braced* so they don't spread with the weight of the concrete (see photo above). Anyone who has worked with concrete can testify to its weight.

Once you've had the terrifying experience of seeing a large mass of wet concrete start moving toward you, you'll always over-brace forms and wire the braces together.

On the inside of the forms, mark a level line for the top of the foundation. A chalk line works well for this. Make the line about 3" above the actual level to which you are going to build the foundation so that it doesn't smear when the wet concrete is being poured. This line is a convenient guide for a level pour (Figure 19). Another way is to chalk line the exact level and height, and drive nails halfway in along the line. This gives an accurate guide.

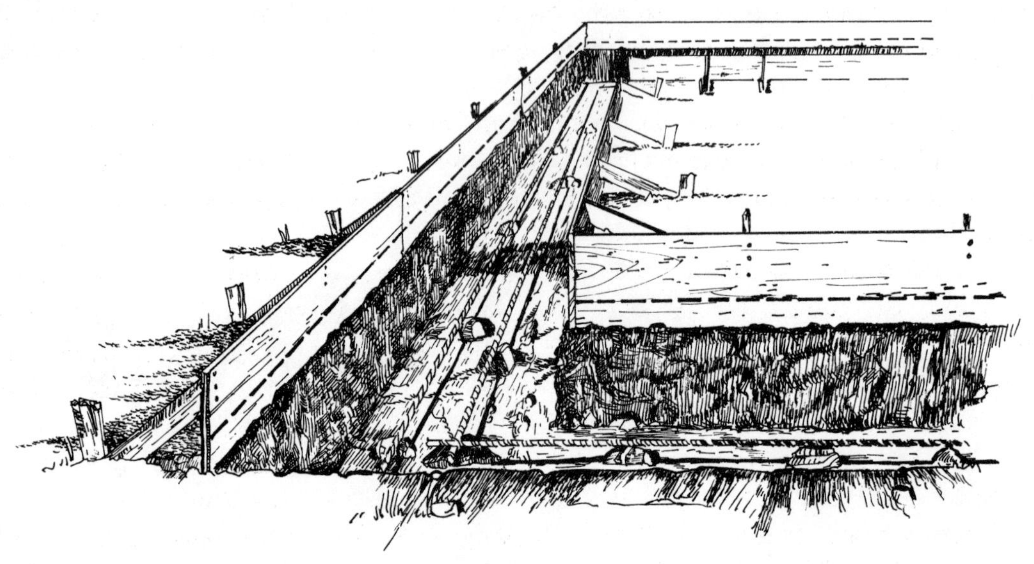

FIGURE 19

The fast way to pour the foundation is to have the ready-mix concrete truck back up to the site and dump it on you. But often the concrete companies won't deliver in small quantities, or the site is impossible to reach. In that case you have the option of buying premixed dry bags or making your own mix from cement and sand. If mixing your own (much cheaper), a standard concrete recipe is five parts sand and ¾" gravel (mixed equally) to two parts dry Portland, and water.

This is a heavy job so line up a few friends. The entire foundation should be poured at one time. You don't want to have seams from two or more separate pourings.

So, you're ready to do it. Sand and gravel in place, Portland bags stacked, shovels in the ready position, wheelbarrow greased, beer iced down. Consistency is what you want in the mix. It should not have dry clumps or an overabundance of any ingredients. The mix should be wet without being runny. If you pull a hoe through it, it should make nasty noises. When the mix is just right, it reacts like Jello when patted with a trowel. Nice stuff.

Start at one end of the trench and work around. After a load, usually a full wheelbarrow, is dumped, spread the concrete along the trench. Push and work the mix down into the trench with a trowel. Don't be gentle. You want to avoid holes or pockets in the foundation. Keep adding loads of concrete until you've nearly reached the level line or marker established as the top of the foundation.

As you work around the trench, pat and smooth out the top of finished areas. After the cement begins to set up, insert an anchor bolt (screw threads exposed) for sections such as low door jambs that will be framed above the foundation. With a square make sure that these bolts are perpendicular and in line with where the plate will be and that you've left about 1½ inches extending above the poured foundation (see Figure 20). The plate, which we will talk about in various contexts throughout this chapter, is not something on which dinner is served. It is a piece of lumber, in this case a 2x4. The foundation plate provides a base for the frame walls of your greenhouse. The top plates give vertical studs and roof rafters something to hang on to. In general plates serve as weight supporting members of any frame structure.

Note: clean your tools immediately after use or they'll never be the same. If for some reason you have to leave a load in the wheelbarrow or mixer for a short time, pour a small amount of water on top of it and cover as tightly as possible. This also applies to mortar and plaster mixes.

After the pouring is done, check to see if any areas have sunk and make sure that the above-grade forms are secure. When the concrete has set up or hardened (usually within three or four hours), spray it with a light mist of water or cover with wet hay or straw. This prohibits rapid evaporation that might crack or weaken the foundation. Spray it every few hours for the next day or two (don't bother at night).

When the foundation has a feeling of permanence, the forms can be removed. Clean well and recycle them into shelves, tables or bed frames for the greenhouse interior.

There is a way to avoid laying a foundation under the frame portions. I'll give it to you as an option. This method is common in large commercial greenhouse construction,

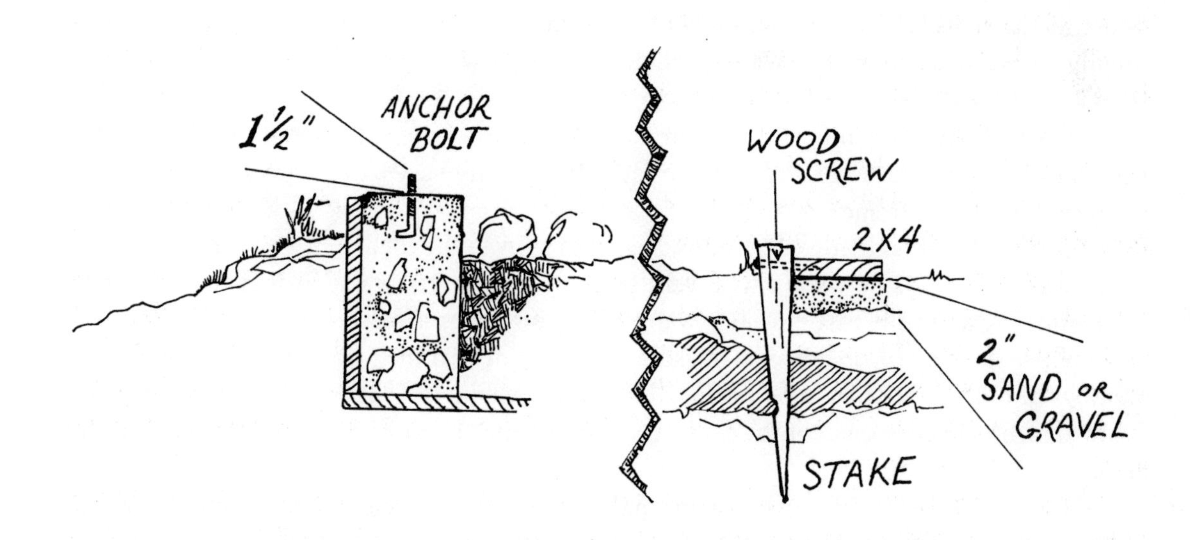

FIGURE 20

and might be useful if you are in a big hurry or plan a temporary structure. Level the ground where the frame walls will be. (Don't dig a trench; just level the earth.) Wood plates (I've used railroad ties) are laid directly on two inches of sand and staked in at 3 to 4 foot intervals. The stakes can be metal or wood but should be at least 36" long. They can be screwed, nailed or bolted to the ground plate. The wood should be treated with copper naphthenate as a preservative. Don't use fresh creosote or pentachlorophenol ("penta"), as these chemicals give off fumes that are noxious to plants. See Figure 20.

Whatever foundation method you choose, the most important considerations are:

1) Is the weight evenly distributed?
2) Is the foundation level?
3) Will the water drain away from it?

If these criteria are met, the foundation will be functional.

MASSIVE WALLS

When the concrete in the foundation has cured, you can begin work on the massive masonry walls. Different types of masonry construction call for different techniques. Let's use hollow pumice blocks for our example. (Basically the same technique applies to building brick or adobe walls, except that you can use mud for mortar in the latter case.) The standard size pumice block is 15½" long x 7½" high x 7½" wide. (They also come in half blocks and about every other size and shape imaginable.) For estimating the amount needed, use the dimensions 16" x 8" x 8" because of the added space to be filled by mortar. Before making your estimation, determine the exact size and location of any vents or doors in the walls. They must have a jamb or frame built around them, and that lumber is usually one and a half inches wide. Include twice that width (both sides) in your calculations.

I like to avoid openings in masonry walls whenever possible. They involve a precision and degree of patience that I often lack. It's usually easier to locate vents and doors in areas that will be frame.

As the diagram in Figure 21 (following page) shows, the high vent is set in the eastern frame wall ("A"). The low southwest vent sits on the masonry wall ("B"). The lowest part of the door is set in the east masonry wall ("C"), the upper 4/5ths in a frame section.

When the size of openings in masonry walls is determined, the estimate of the total number of blocks needed can be made. Determine the square footage of the walls and estimate one block per square foot plus 10% for cutting. For our example I'd buy 80 full blocks and 25 half blocks.

However, if you're having problems making the necessary calculations, remember that elderly, experienced salesperson we mentioned earlier. Write the dimensions of the walls (with doors and vents figured in) on a piece of paper and go to your local hardware store. Odds are you will get all the help you need and some good advice on the side.

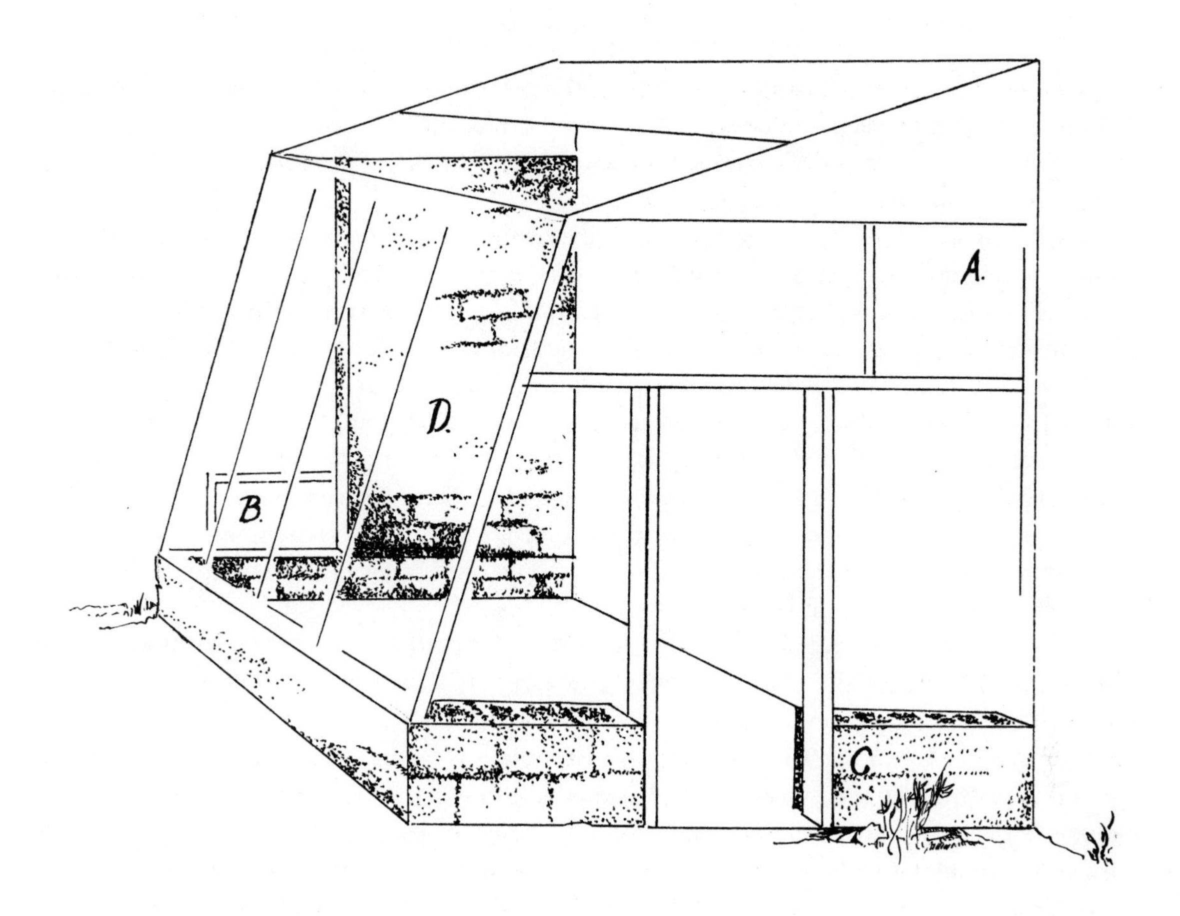

FIGURE 21

You may want to order more than we've estimated. Masonry blocks look rather formidable, but are actually quite fragile and easily broken until they're in the wall. Handle and stack them carefully. If you do have leftovers, most reputable building supply stores will refund money on items returned in good condition with a receipt.

The mortar used in laying blocks is made with masonry cement and screened sand. The standard proportions are 5½ parts screened sand : 2 parts masonry : 1 part Portland. Mixing equipment is the same as for the foundation. Small amounts are made as needed. A large triangular trowel is used to spread the mortar.

To lay a perfectly straight and level block wall, poles can be erected precisely at the outside edge of each corner of the structure. Marks are made up the poles at 8" vertical intervals. Poles and marks are checked against each other with heavy twine and a string level. They should all be at exactly the same level for each course of blocks, and they indicate the top of each layer. That's the precise way to do it.

Another method is to simply begin laying the blocks, checking for level, plumb, and straightness as you go. In an area as small as our 10 x 16' example, this should be sufficient. Lay the first course of blocks all the way around. Check for square on the corners. Now, using the level, build up several courses of block at each corner. String is strung between corners and straight runs laid down to the string. Checks for vertical can also be made with a carpenter's level or plumb line. If you're new at this kind of work, don't trust your eye too much.

Have you ever watched an experienced mason building a wall and seen the beautiful fluid movements he makes? A quick scoop of the trowel places a glob of mortar on the edge. The mason flicks his wrist downward, a microscopic spasm, and the mortar adheres to the blade in a flat, compressed mass. A long thin line of cement is spread along a four-foot edge of the existing wall in the next stroke. Another similar motion, backhanded, lays it on the other edge of the wall. The third trowel of mortar is applied in fast choppy slaps to the center ridges. The new block is picked up and hit with mortar on two vertical edges, and wham, in it goes. This takes about six seconds. Like a finely tuned human machine, the mason progresses along the wall with the speed, economy of movement and accuracy found in downhill racers and basketball centers. The only thing that slows down a pro like this are scaffold movers and unions.

Don't expect to match this degree of skill. Try to put a uniform thickness of mortar (3/8 to 1/2 inch) on all seams. Any way you can get it to stick to the blocks is cricket. Try putting a small amount on the trowel and scraping it off with that downward and outward motion. If you can't get the knack of this, put a larger amount on the trowel and shake it over the edges. I resort to my hands occasionally. "Seat" (firmly tap down) the block with the handle of the trowel. It should be evenly supported by mortar and level when it's in place. See that the seams are staggered, not one directly over another.

A method used to "tie-in" a new masonry wall to an existing structure is to bend a small (8 x 10") piece of metal lath to form a 90° angle and fit this between the new wall and the home wall. Tack the lath securely to both walls. Lay the mortar and firmly seat the next course on top of the lath, pushing the block tightly to the existing wall. This should be done three or four times in a wall as tall as the one in our example, the 8' west wall.

When the block walls are up to their final height, fill all center holes in the blocks with concrete (the same mixture you used for pouring the foundation). This strengthens the walls and adds mass, heat storage capability, to the structure. Before the concrete dries, insert anchor bolts where the framing plates will be applied to the top and sides of the walls (see Figure 20).

When all the masonry walls are up, relatively straight, level and plumb, it's a good time to have a celebration. You can finally see and feel the results of your brain and muscle work. The hardest part is over. Enjoy it!

FRAME WALLS (CLEAR AND OPAQUE)

As stated earlier, a frame wall does not have any appreciable mass; therefore it cannot store heat for a structure. However, when properly insulated, it will keep the heat in and it is often justified if other heat storage is planned. In the diagram in Figure 23, page 39, we show frame walls for purposes of illustration.

The diagram also indicates that the vertical framing members (studs) in the *clear-wall* sections are erected at 47 inch intervals (measured in what's called centers. Centers are determined by measuring from *one* edge to the *next* edge on the *same* side.) The fiberglass that we recommend comes in 48" widths. The 47" center allows for overlapping the panels. Studs in the *solid frame* wall are on 24" centers, as are the roof rafters. All corner studs and upper plates are double-width for added strength. A scale drawing of your greenhouse using this type of specifications will help you estimate how many 2x4's of various lengths will be needed for the rough framing. Add the total linear footage of doors, vents and horizontal nailers (firestops) to the estimate. Add another 20%. We've supplied a list for this 10 x 16 foot unit at the end of the chapter.

You'll save money by having as few leftover scraps as possible. This is accomplished by making all principal framing members slightly shorter than an even number. Hence, if the stud in the south wall is 7' 10" high, it can be cut from an 8' piece. On the other hand, if the stud is 8' 1" high, a 10' 2x4 usually must be purchased, and you're left with a 23" scrap. At current lumber prices, that's sinful. In your scale diagram, *make* the lengths come out economically. Do this by slightly changing angles and dimensions in the drawing until it works. If you just can't make it come out right, then plan to use the scrap lumber for tables, shelves, boxes, bed frames or other things. A 23" scrap, for instance, could be used for a firestop in the sheathed (solid frame) wall.

It's important to get a proper dollar value for your lumber. As with most things in life, you do this by choosing it yourself. Service personnel in the lumber yard are usually happy to let you choose and load your order. Occasionally it will be "company policy" not to let the customer look through the stock. Don't do business with a company like that.

Hold up each stud and look down the entire length for straightness. It should not be warped, twisted or badly bent. A slight curve is to be expected; it is called the "crown." If you are holding the stud on its edge, you can easily see the crown. The crown is not to be confused with a bow in the broadside of the lumber which, if it's slight, can be taken out in installation. In construction, the crowns should all face the same way on the walls and should always be arch-up on the roof.

Next, check the stud for clearness. Does it have an unusual number of knots? Reject it. Does it have heavy "pitch" or "sap" areas? Reject it. Soundness can be tested by tapping it against a solid object. The stud should sound firm and full, not tinny or dead. When making all these tests, carefully place the rejects back on the pile or the yard people will never let you do it again. Lately, I get about a one-out-of-five acceptable ratio (that should give some idea of how much lumber you're moving around). Stud lumber, No. 2 common grade, won't be perfect, but get it as straight, clear and dry as you can.

After all this trouble in selecting the lumber, be sure to stack it flat and keep it dry when you get it home.

FIGURE 22 PHOTO BY KRISTEN MACKENZIE

Let's begin by framing up the front face. First, you have the option of framing the wall in place or building it "prefab" on the ground (see Figure 22, above). This decision is mostly a matter of personal preference. We're going to prefab the front face in our example, and frame the side walls in place.

In the diagram on the following page, Figure 23, the length of the roof and foundation plates (top and bottom framing members) is exactly 16' (No. 1). The top plate should be double to prevent bowing under the weight of the roof. So the first step is to nail the two 2x4's together with 16-penny nails. Place the 2x4's flush and drive the nails in at an angle so that they don't protrude from the wood.

Lay the plates (bottom and top) on the ground and mark them where the vertical studs (47" centers) will be attached. Corners should be double-width (No. 2); use 4x4's or sandwich two 2x4's together with 3/8" plywood or wood lath in between to bring the dimensions to 4" x 4".

The next step is to cut the lumber to the appropriate length. In our example, we have seven studs cut to 7' 7½". With the plates, this makes an 8' high front face.

Now if the south wall is to be vertical, the cuts are square on each end of the studs. In our example, we have tilted the front face twenty degrees to make a seventy degree slope. (With any tilted south face, the bottom angle to the foundation plate will be the difference between the south face angle and ninety degrees.) All we need to do here is mark off twenty degrees from a ninety-degree cut. Make the top cut parallel to the bottom so that you have a horizontal surface for the top plate. When the studs are cut, nail them at the marks to the top and bottom plates with 16-penny nails. Note: always buy high quality nails.

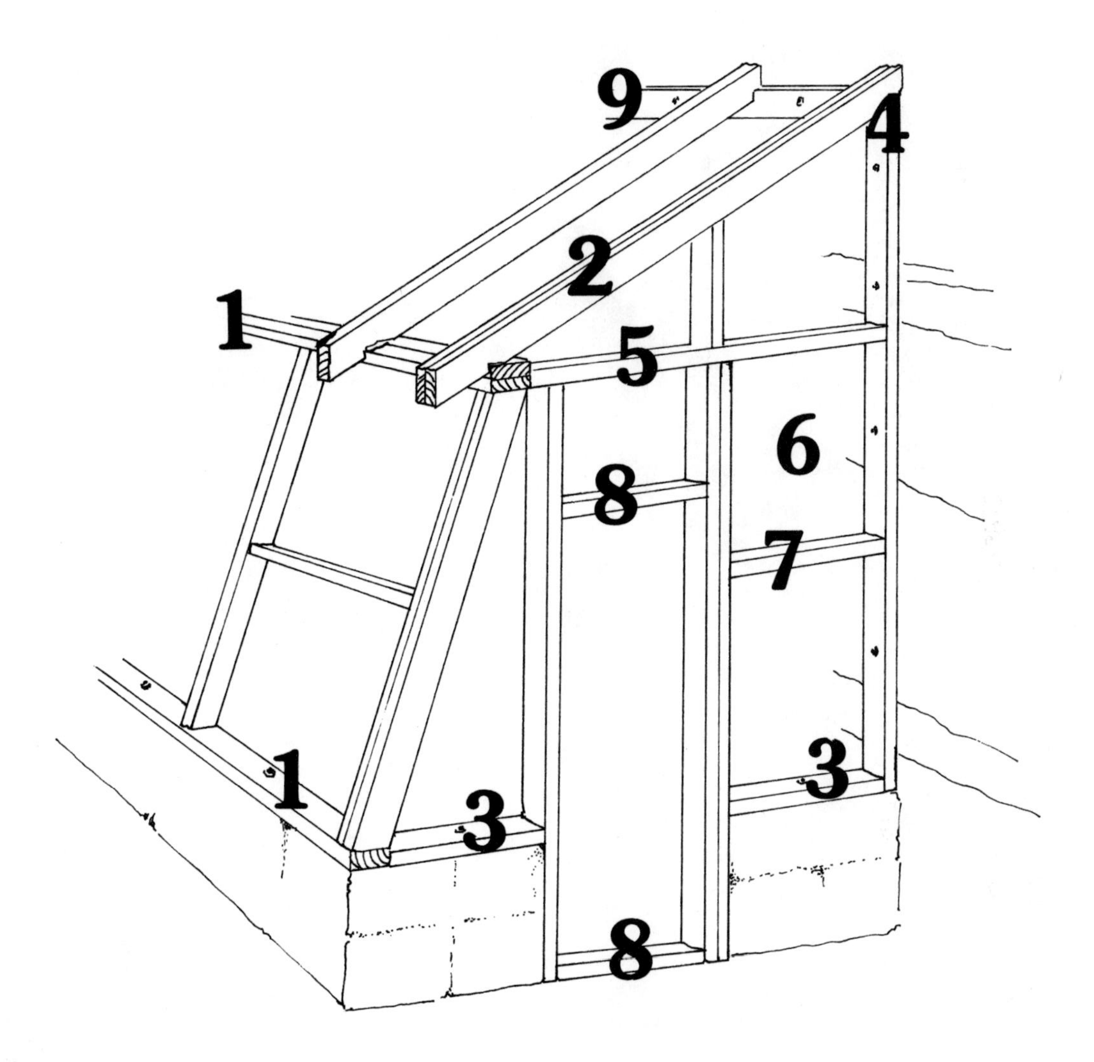

FIGURE 23

The south panel can now be lifted into place on the low front wall. This will take several folks working in unison. In order to precisely mark the junctions of the anchor bolts in the low wall with the bottom plate, gently lower the prefab frame structure into its permanent position. While several people hold the framework in place, one particularly aggressive type can pound on the plate with a hammer directly over the protruding anchor bolts to mark their position. Take the face off again and drill holes at the indicated marks. The front face can then be installed on the wall, temporarily braced and bolted down.

At this point put up a few rafters (No. 2) for braces, the ones on the corners and a couple of others. (Also, see Figure 24 on following page.) Nail them down to the top plate of the front face and to the plate or rafters on the house side (see next page for instructions). This will give the structure stability and give you a chance to see the outline of the greenhouse. When you have finished attaching a few rafters, take a break. Stand back and admire your work.

FIGURE 24

The next step is to cut and place the lower plates on the east and west walls (No. 3, Figure 23). Bolt the plates down. To connect the frame walls of the greenhouse to the home, a stud is securely "tied-in" to the home wall. It should fit snugly from the bottom plate to directly under the rafter plate (No. 4).

You now have a funny shaped "box"; that is, the perimeter of the east and west walls to fill in with framing lumber. We're going to divide that box with a plate that will also serve as a "header" for the lower frame walls (No. 5). On the east side the plate can also be the top of the door frame. Hold a piece of lumber level across the span and mark its intersection with the front face. On the west, it sits on the massive wall for the majority of the span. When the pieces are cut, nail them into place.

The east and west walls can now be framed in place. Remember, the clear sections will have 47" centers. The insulated walls will be framed on 24" centers (No. 6). The easiest way to do this is to mark the bottom plate at the intervals where the upright studs are to be nailed. Take a carpenter's level and hold it against the side of the first stud. Keep the base of the lumber on the bottom mark and get the stud exactly plumb. Then mark the point of intersection of the stud with the header on both the stud and the plate. Cut the stud about 1/16" longer than the mark you've made.

All frame walls can be constructed in this way. It's easy and it's fast. Toenail (drive nails in at an angle) the studs in as you go. When the studs are cut, fitted, and nailed, install horizontal spacers (No. 7), and the vent and door plates (No. 8). Make certain they're level, and nail them in with 16-penny nails.

ROOF RAFTERS

With luck, you will be able to tie the roof rafters of the greenhouse into the existing rafters of your home. If not, attach a 2x6" or 2x8" plate to the wall of the house as a base for the rafters (No. 9). Expansion bolts or large wood screws are used to get a *secure* tie-in to the wall. Don't scrimp here. The greenhouse roof must bear its own weight plus, in many areas of the country, snowloads *and* the additional loads of snow sliding off the roof of your house.

The rafters for a short span (under 10 feet) can be 2x4's or 2x6's set on 24" centers. For longer spans use heavier lumber or put the rafters on closer centers.

Shallow notches are cut in the wall plate or joist hangers can be used instead. The rafters are toenailed to it at these points. At the intersection of the rafters and the south face top plate, very shallow notches (called "bird's mouth" notches) are cut in the bottom of the rafters so that they will rest snugly on the plate. Again, toenail them in.

Putting up the rafters (and the roof) can force a person into some pretty strange acrobatic contortions. I would caution you that eight to nine feet up in the air is higher than you might imagine, especially from an aerial view. A body can be broken at the end of a free-fall from that height. Also, get in the habit of *not* leaving any tools or materials lying about on the rafters or the roof (even for a moment or two).

When all the rafters are in place, mark the intersection of the clear roof with the insulated roof areas. Use a chalk line to do this (see Figure 28, page 46). To determine that junction in your unit, use the charts as explained in Appendix A. It will usually be about halfway down the rafter span.

PAINTING THE FRAME

Rafters and framing lumber can be treated in more humid climates. Copper naphthenate is recommended. Some paints have a preservative in them. After it's been treated, all

FIGURE 25

framing lumber that will be visible within the clear walls and ceiling areas should be painted with a glossy white enamel or latex (see Figure 25). This will help reflect light into the greenhouse and also enhance its appearance.

CHOOSING THE CLEAR GLAZING

Traditionally, glass has been used for the clear surfaces in greenhouses. Due to its resistance to high temperatures, glass is also used extensively for covering solar collectors. Prefabricated sheets of double-layered glass are available from major manufacturers and may be purchased in various sizes. Many types of sliding glass doors are also double glazed and can be used in the greenhouse. The obvious advantages of glass are that it allows a view to the outside and is highly resistant to the harmful effects of weathering. It is, however, easily broken and demands extreme care and technical skill in installation.

Technological advances in the design and manufacture of plastics have produced important alternatives for the greenhouse builder. Some types of semi-rigid fiberglass/acrylic sheeting are guaranteed for 20 years to transmit enough light for photosynthesis. However, fiberglass can become cloudy or brown as a result of ultraviolet ray damage ("blossoming"). Tedlar coating, for instance, has an ultraviolet retarding characteristic that helps greatly to preserve the clarity of the plastic to which it is applied.

Fiberglass transmits nearly the same amount of light as glass, even though it's translucent rather than clear (see Figure 26). Corrugated plastic is recommended for clear roof areas; it is easily installed and resistant to hail damage. Flat fiberglass is readily attached to clear-wall frames with rubber-gasketed nails or lath strips. I recommend the use of flat fiberglass on all vertical and near vertical surfaces. Flat material has a 20% smaller surface area than corrugated; therefore far less area for heat loss. The quality of light transmitted through fiberglass is diffuse. It doesn't give the sharp, clearly defined shadow areas of glass. This is beneficial to plant growth. The cost per square foot of new fiberglass is considerably less than the cost of new glass.

FIGURE 26 PHOTO BY KRISTEN MACKENZIE

There is an ecological question concerning the use of plastics in general. Plastic is a petrochemical product and is not biodegradable. The supply of petrochemicals is dwindling rapidly, and the atmosphere is becoming polluted by petrochemical wastes. Nondegradable products also constitute a form of pollution on the earth.

I feel that our only hope for maintaining an ecological balance on the earth depends on a thoughtful, positive use of modern technology and its products. I believe that employing fiberglass and plastics in your greenhouse constitutes such a use. Many people are convinced that solar energy is the obvious fuel of the future. Further experimentation will lead to advances in the efficiency of solar collection, storage and transmission. It is probable that plastic products will become essential to the production of solar energy components. I suggest, therefore, that the intelligent and careful use of plastics in the greenhouse will serve as a valuable example of how to put our fossil fuels to beneficial use...before they are exhausted.

CLEAR WALLS

Having built and painted the frame walls, you are now ready to install the outer layer of fiberglass. Most of the products (Lascolite and Filon, for example, brands we've used and can recommend highly) are sold in 4 foot wide rolls, up to 50 feet long. For this reason, we suggest that you install one 4 foot panel at a time. Beginning at one end of the frame wall, measure the height of the section carefully. After measuring and marking the material with a felt-tipped pen or razor knife, cut the section with sharp wire-cutting shears. Apply an even bead of sealant, silicone or butyl rubber to the wood frame you are going to cover. With two or three people holding the edges of the plastic, align it in front of the frame. It is handy to make corresponding marks on the fiberglass and the wood frame to aid in alignment. Nail the top edge to the upper frame member in the center. Use a gasketed nail; do *not* nail it in completely. Check to see that the sides and bottom will fit properly. Nail the outside edge in the center (again, halfway in). Pulling slightly on the remaining two edges (bottom and inside), nail them temporarily in the centers of the frame. You can use flat-headed 1" nails for the inside or leading edge or simply lay it on with no nails. The next section of plastic will overlay it and must fig snugly. The rubber-gasketed nail heads protrude about 1/8" from the wood and would cause bumps in the overlaid plastic. Nail the material down to the horizontal braces first. Then pulling diagonally on the four corners of the sheet, put in nails approximately 8 inches apart and angle them out to pull tension on the sheet. Work the bulges out of the plastic from the center to the corners (see Figure 27 on next page). If a major bulge has developed, try to detect it early; remove the temporary nails and realign the sheet.

There are two very expensive items used in this method, silicone sealant and gasketed aluminum nails. Lately I've used a cheaper method that I believe is just as effective. Check and mount as before. Don't use any silicone sealant. Get small galvanized nails and drive them into the fiberglass about every six inches. After all panels are up, cover edges and overlaps with thin (¼" x 1½") wood lath. If there are any bulges or leaks, they can be sealed with regular caulk.

After cutting the second panel, lay another bead of adhesive over the leading edge and the top and bottom. Proceed as above, overlapping the second panel by 1 inch on the edge previously flat-nailed. Use nails to attach the overlapping section. Follow the steps given for the first panel, again using 1" flat-headed nails in the leading edge to be overlaid by the next panel. For vents and removeable panels, install the fiberglass on the ground.

FIGURE 27 PHOTO BY KRISTEN MACKENZIE

The irregular shaped panels in the corners of the east and west wall are sized by cutting a rectangular piece to fit the length of the frame. Hold the piece up to the panel and mark the shape with a chalk line. Save the scraps for other irregular shapes and vents.

ROOF INSTALLATION

Spanners (No. 7 in Figure No. 23, page 39) like those used between the vertical wall studs will be nailed between the roof rafters to receive the corrugated plastic and solid roofing material. Be sure to toenail-in a straight row of these spanners along the line separating the clear from the solid roof (Figure 28, page 46). In all other roof areas, the spanners can be staggered for ease of nailing (Figure 29, page 46). They should be spaced up and down the rafters at three to four foot intervals.

You are now ready to install the corrugated clear roofing. One sheet of corrugated is held in place at a corner. Align it carefully along the bottom and side edges. The roof can overhang the side two corrugations and the bottom about 6" unsupported. Get it "true," that is, perfectly in line, with the rafters and plates. An out of line first piece will throw the roof whacky. When you've got it right, tack it down temporarily at the top. On the bot-

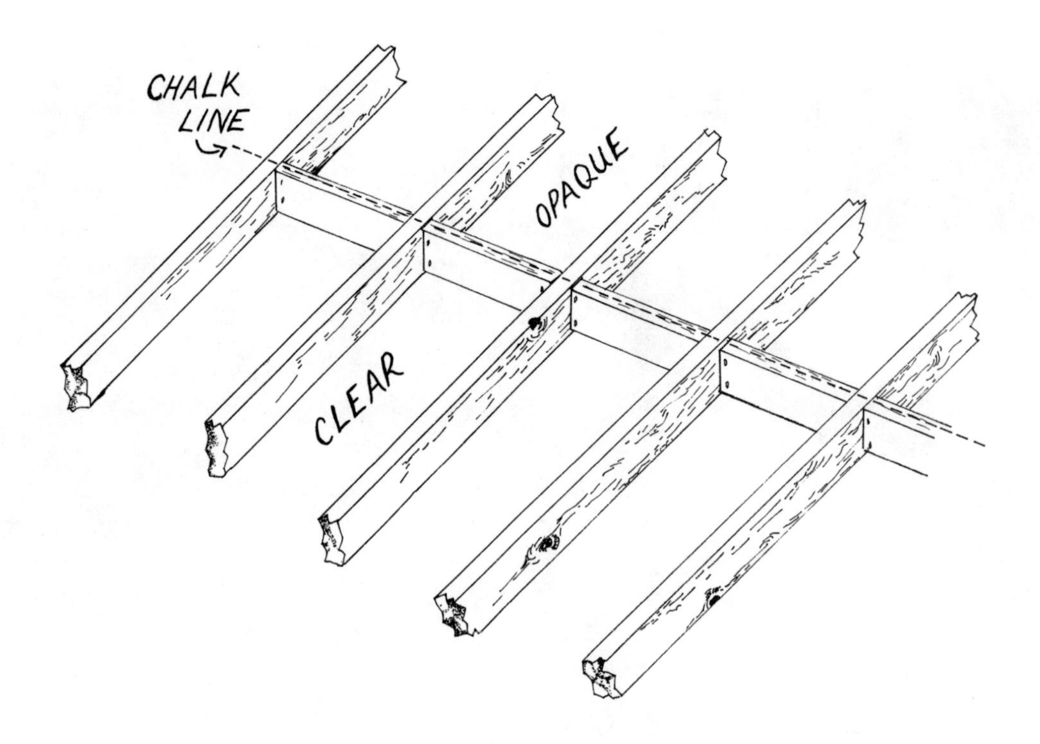

FIGURE 28

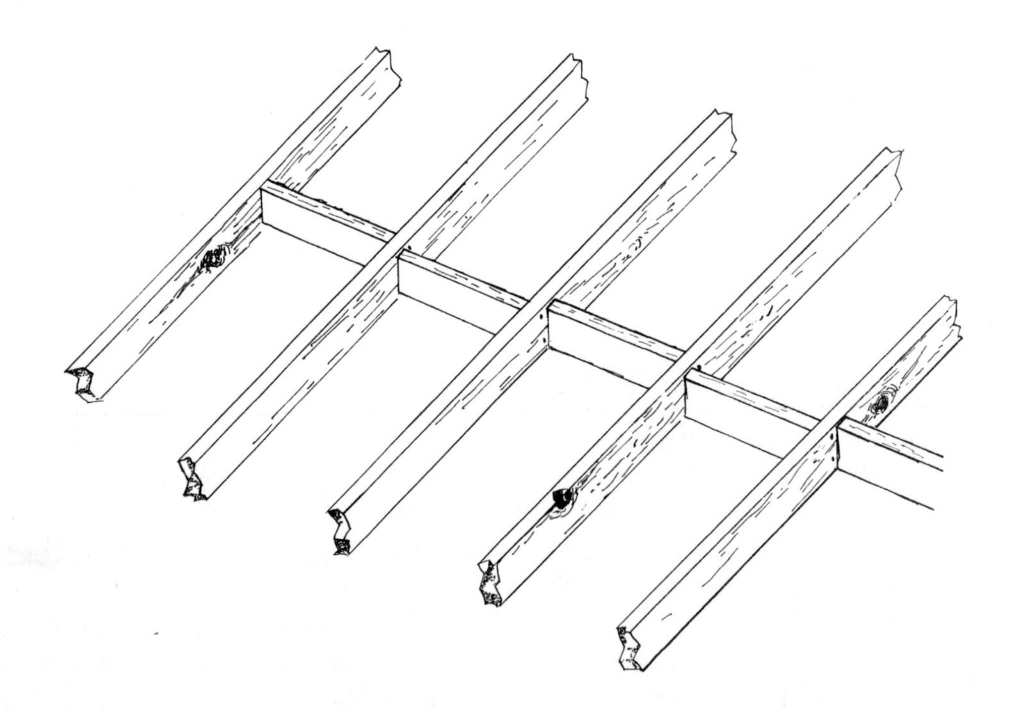

FIGURE 29

tom plate, under the fiberglass, insert a corrugated sealing strip. I recommend the foam type, but redwood or rubber will do fine.

Now nail down the fiberglass every 3 to 4 corrugations with gasketed nails to the bottom plate and rafters. Nail into the high ridge rather than the valley. This will prevent water seepage through the nail hole. Overlap the sheets at least one and, if possible, two corrugations and continue across the clear area in this manner. Don't nail the top edge (to be overlapped by the opaque roof) yet. The only real trick to nailing into fiberglass is to have a steady hand and a good eye. A missed hammer blow can splinter the material, leaving an ugly opaque mark (to say nothing of the damage to your thumb).

The solid roof is installed next. If plywood and composition roofing material is to be used, lay a strip of corrugated molding across the top edge of the clear/solid roof junction. Nail the plywood sheets onto the rafters, overlapping the foam strip and the clear roof area by 3 to 4 inches. The foam molding above the corrugated plastic will create a tight seal against heat loss. You can now apply your composition shingles or other roofing material over the plywood.

Vents in the roof are *not* recommended for any but the most experienced builder. It is extremely difficult to seal them against air and water leaks. If you have the knowledge, build a "boat hatch" type vent of 1x4's that extends about 2" above the roof. Fit the vent so that it lays flush on the 1x4 box.

In New Mexico a common roofing material is corrugated galvanized steel. It is relatively cheap, easily installed and maintenance free. If you use metal, no plywood is needed. Simply align the corrugations of the galvanized with those of the plastic (they match), and nail it down to the rafters and spanners. Again, the solid roof should overlap the clear areas. For nailing, hold lead-headed roofing nails with a pair of pliers over the designated spot (on a ridge and over a rafter or spanner) and strike solidly with a hammer. Wear eye protection.

THE INTERIOR PLASTIC

Six-millimeter polyethylene (with an ultraviolet inhibitor) is suggested for the interior clear walls and roof areas. It is susceptible to weathering but will be protected by the fiberglass outside layer; it should last three to five years before having to be replaced. I've used both Monsanto 602 and Tedlar.

The thin plastic is more easily installed than the outside fiber acrylic. Large sections of clear area can be covered at once. Using scissors or a razor knife, cut out the sections to be attached. Make your cuts at least 3 inches longer than measured. With three or four helpers holding the extremities of the sheet, begin stapling along the centermost stud. Work outward toward the sides and corners, stretching the sheet as staples are driven 8 to 10 inches apart. Wood lath (¼" x 1½") is nailed over the plastic-sheeted studs with finishing nails to produce the final seal. You can stain or paint the stripping to make it more attractive.

INSULATING FRAME WALLS

Solid frame walls can be insulated with a wide variety of materials. Rockwool, fiberglas, styrofoam, polyurethane, pumice and cork are all excellent. Foil-backed insulation stops a lot of radiant heat loss through the walls and acts as a vapor barrier, but it can be prohibitively expensive. If you do use it, install it with the foil backing facing the interior of the greenhouse. Recently I've been stapling heavy duty tin foil on the interior side of fiberglas batt. That's cheap and effective.

The amount of insulation applied should be as much or more than is used in the walls and roof of a well-built home in your area. In New Mexico, I usually use at least 4 inches of fiberglas batt in the walls and 6 inches in the roof. Fiberglas is easily applied with a staple gun or tack hammer. Salespeople at the local hardware store can help you choose the right amount and type of insulation.

Always wear a long-sleeved shirt, gloves and button your collar when installing fiberglas insulation. Wear safety glasses for rafter work, especially if you have sensitive eyes, and try not to breathe too much.

INTERIOR PANELING

Insulated frame walls can be paneled with any material that you find attractive. Masonite, plywood, sheetrock and rough or finished lumber are widely used. Paneling materials call for various adhering techniques. Wood products are usually simply nailed with finishing nails. Some paneling can be glued with construction adhesives. When using large panels, measure carefully before cutting. If you're careful, you can save odd-shaped scraps for other areas and keep waste to a minimum. Occasionally you may have to nail in a "scab," a scrap of 2x4, to the stud walls in order to give a nailing surface for the paneling. If corners don't "flush out" perfectly, don't worry. That's what God made corner lath for.

As the insulated wall will not store heat, the inside should be a light color to produce a reflective surface. Water-sealing will help to protect the interior against deterioration due to high humidity.

DOORS AND VENTS

Doors and vents must be tight fitting and weather-stripped around all joining surfaces. If you are using fiberglass, you may choose to use glass for the vents and/or doors. Clear views to the outside can offer attractive accents to opaque and translucent walls.

When glass is used, it should be double glazed to reduce heat loss. If possible, windows and doors should slide or hinge away from the greenhouse interior. Hinges are mounted on top of vents to open out. Remember, the high vent that you put in one wall (east or west, depending on the direction of the prevailing winds–should be on the downwind side of the prevailing air flow) needs to be about one third larger in square footage than the low vent in the facing wall.

In determining the dimensions for an outside door, remember that you will be moving large quantities of soil into the greenhouse. Make the door wide enough to accommodate a wheelbarrow (plus your knuckles). Thirty-two inches is good. It is also advisable to make the height of the door a standard measurement for convenient access. In very cold climates an air lock over the exterior door will save large losses (see Herb Shop, page 87).

EXTERIOR PANELING AND INSULATING

All massive walls should be insulated on the outside. An effective insulating material for this is styrofoam or styrene panels (1 or 2"). They can be stuck to the walls with a heavy duty construction adhesive. Use it liberally. If the wall is to be plastered, cover the styrofoam with tar paper. Then use firring nails and chicken wire over that. The wall can now be plastered with a hard coat (5 parts sand : 3 parts Portland : 1 part lime).

Exterior covering of the frame walls can be any material that suits your aesthetic and economic criteria. I've used old lumber, plywood, Celotex (exterior fiber sheathing) and metal siding. The most important thing is to make sure there are no leaks that allow water into the frame walls. Remember–higher panels overlap lower ones for waterproofing. Caulk anything that looks suspicious.

SEALING THE GREENHOUSE

If there is one most critical factor in the construction of your solar greenhouse, it is that *all joining surfaces fit tightly together.* Tightness is, in fact, one of the highest goals which you can attain, both as designer-builder and proud owner. The primary reason to strive for tightness is to reduce air leaks. But another important reason is to produce an attractive environment that reflects the care and thoughtful work that went into the design and construction.

During each phase of building, ask yourself where heat loss is likely to occur. Wherever large surfaces are joined (such as the foundation plate to the foundation), a layer of fiberglas or foam insulation will help solve the problem. Sealants such as silicone and caulking materials (wood filler, spackling compound) should be used on any crack or opening that could transmit air through the structure. For larger openings, such as might occur at the house/greenhouse junction, use metal lath and plaster to build air-tight walls (insulate between them). As we mentioned earlier, vents and doors must be completely weather-tight. The importance of sealing and insulation cannot be overemphasized, as it can make the difference between the success or failure of your greenhouse.

THERE YOU ARE! YOU DID IT! CONGRATULATIONS!

TOOLS NEEDED

Shovels (pointed and flat)
Hammers
Saws (hand) or
Saws (electric)
Drill and bits
Hoe
Trowels (triangular and plaster)
Hatchet
Level
Tape measure
Squares (tri and carpenter's)
String
Draw knife
Staple gun
Saw horses
Paint brushes (3 inch and 1½ inch)
Wheelbarrow
Rasp
Plane
Tin snips (large)
Caulk gun
Chalk line
Roller and pan
Water buckets or hose and nozzle

MATERIALS LIST FOR 10' x 16' ATTACHED SOLAR GREENHOUSE

1 bag masonry
5 bags Portland
1 yard sand
1 yard ¾" aggregate
2 yards dry pumice or pea gravel
80 full pumice blocks
25 half pumice blocks
8 6" anchor bolts
72' rebar
30 8' 2x4's
3 10' 2x4's
2 16' 2x4's
2 8' 1x4's
2 10' 1x4's
40' 1x12 (for floor beds and shelving)
400' wood lattice moulding (for trim and tables)
½ lb. concrete nails
300 - 400 aluminum nails or 3 lbs. small galvanized nails
10 lbs. No. 16 common nails
5 lbs. No. 8 common nails
2 lbs. No. 8 finishing nails
1 lb. small finishing nails
3 lbs. blue sheetrock nails (for sheetrock)
6 sets hooks and eyes
1 set 3½" or 4" butt hinges (for door)
2 sets 2" butt hinges (for vents)
7 2x4 joist hangers
2 4x4 joist hangers
3 pulls
8 corner braces (to reinforce door and larger vent)
2 2'x8'x2" styrofoam panels
150 sq. ft. of 4" or 6" fiberglas insulation 24" wide
2 packages 3/8" foam strip (weather-stripping
32' corrugated stripping (foam or redwood)
4 pieces ¼" or 3/8" sheetrock
2 pieces ¾" Celotex or equivalent exterior sheathing or paneling, i.e., 64 sq. ft. rough lumber
4 pieces 8' corrugated roofing material
2 tubes silicone caulk–clear
1 tube regular caulk
1 gallon good quality white latex paint
1 gallon good quality dark color latex
1 pint dark stain (for lattice moulding)
200 sq. ft. flat fiberglass/acrylic (greenhouse quality)
70 sq. ft. corrugated fiberglass/acrylic (greenhouse quality)
250 sq. ft. polyethylene (greenhouse quality)
6 55 gallon drums with tops (water tight) and/or a number of smaller water tight containers

NOTES

CHAPTER V Performance and Improvements

THE BASIC UNIT

The attached solar greenhouse we have described will extend your growing season a great deal without supplemental heat or further modifications. The following graph reflects the extended growing season in such a unit over a year's period.

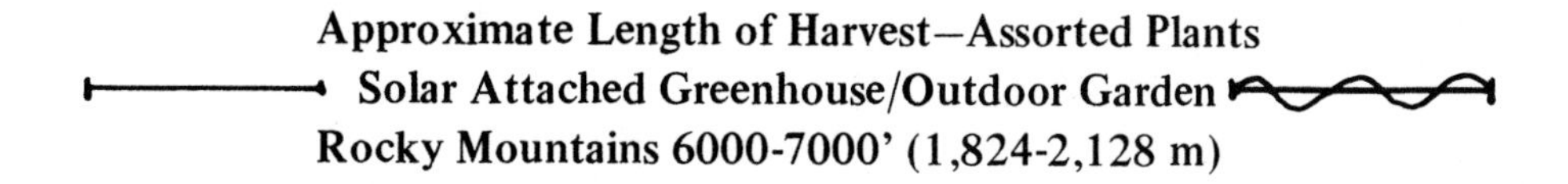

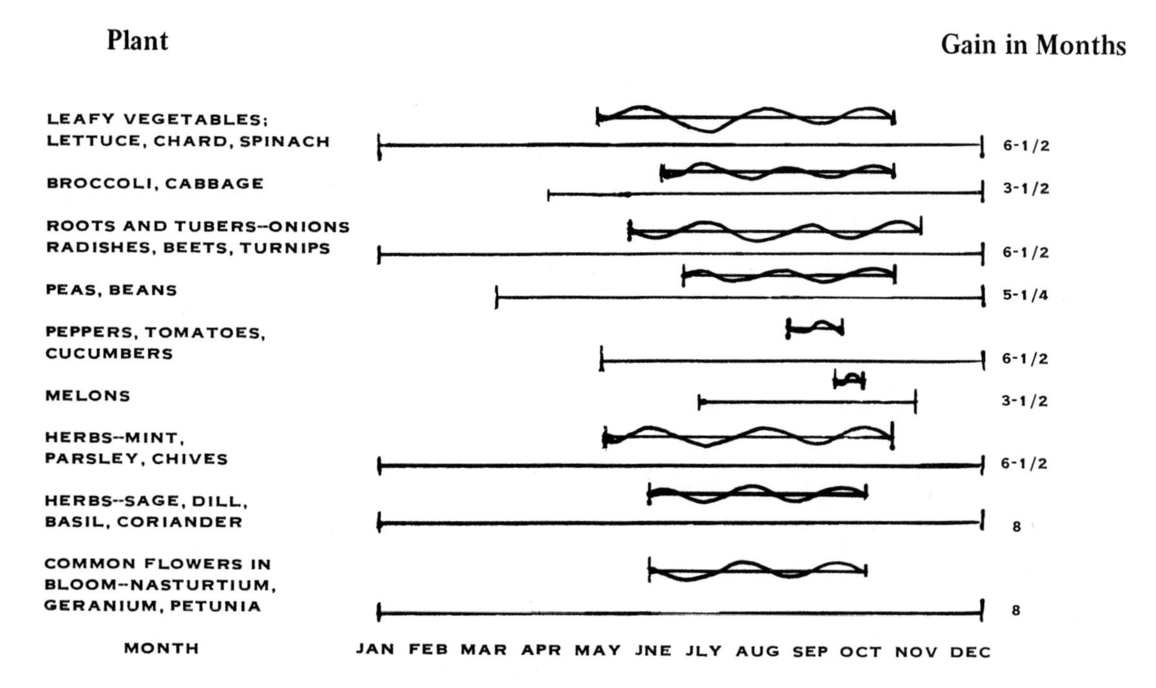

There's an easy way to estimate expected temperatures in your region. Most local weather stations maintain charts of average temperature ranges, sunlight days, rainfall and so forth for their areas. Get the information. Figure that the passive attached greenhouse with the recommended storage will average about twenty to thirty degrees higher than the outside temperatures. This is a ballpark figure, without guarantee; we can't control how tightly you seal your unit, how much insulation you use, or even which way you point the thing. But this is a fair estimation of its potential if you have followed the design, building and operation instructions carefully.

It's a good idea to test the performance of your basic passive structure throughout one year's operation. Chart the highs and lows (indoor and outdoor); record planting, germinations and harvest times. These data will help you determine whether improvements and modifications are needed.

I should mention here that many people prefer *not* to shoot for year-round greenhouse operation. Besides giving yourself a break from fulltime upkeep of the unit, a dormant period offers other advantages. The bedding soil can be composted, fertilized and given a rest. You can also get basic maintenance on the unit done during this time, and plan for the next growing season. If you have had a major insect disaster in the fall/early winter cycle, the rest period can be a good time to freeze the little parasites out of existence (open doors and vents to the outside). On sunny winter days you can still tap all greenhouse heat for home use, then cut it off at night.

The most common reason for shutting down during the four to six coldest weeks of the winter is an economic one. The amount of produce that you harvest during this time may not justify the cost of production, either in terms of conventional supplementary heating or the cost of an active solar heating system.

If your goal *is* year-round production, or if performance improvements are definitely called for (your greenhouse freezes in July), you have three alternatives:

1) Improve collection and heat storage,
2) Decrease heat loss/increase insulation,
3) Employ conventional supplementary heating or open the door to your home.

We like the first two approaches, as they can be employed without tapping conventional power sources. In some cases a limited amount of electricity can be justified, but much less is used than in conventional heating. Solar researchers have pursued two basic directions in improving performance. They are *passive* and *active* systems.

Passive systems emphasize the most effective relationship between collecting surfaces, potential storage mass within the structure, and the unit's insulative characteristics. In a "pure" passive mode there is no machinery making the whole system go. The home or greenhouse is the collector, insulator and heat storer.

In a recent paper published by the United States Energy Research and Development Administration, J. Douglas Balcomb and James C. Hedstrom of Los Alamos, New Mexico, clearly state the importance of passive solar energy utilization. In their words:

> Preliminary findings presented in this paper indicate that it is reasonable to expect comparable solar heating performance for a passive solar building as for an active solar heating system for the same collection area. This is a robust result. The implications to the future development of solar energy for heating buildings could be profound... A number of examples of passive solar heating concepts have been built into structures and received widespread attention. Most use extensive south exposure glass to admit low-angle winter sunshine to the

building, and extensive use of mass heat capacity inside the thermal envelope of the building to store heat energy. Despite the publicity given to these projects and their apparent success in saving energy, they haven't been widely adopted.

We feel that a great deal of emphasis should be placed on the design of passive systems. For this reason we suggest that you make all possible passive improvements before venturing into an active solar system.

INCREASING COLLECTION

It's conceivable, for instance, that you haven't provided enough collecting (clear) walls for your particular conditions. You may discover that the clear roof area needs expanding to allow the sun to hit well up in the back wall during December. Or you might have an east-facing frame wall that could collect enough morning sun to justify its being completely clear. It's not too big a job to convert all or part of such an insulated frame wall to double glazed plastic. Just remove the inner and outer paneling and insulation, paint the studs white, then stretch your coverings as described in Chapter IV.

FIGURE 30

Another way to increase collection is to bounce or reflect more light into the unit. A reflective panel hinged off of the bottom of the front face can also act as an insulating cover (see Figure 30). Hinging reflectors off the sides or roof can also increase collection, but the damage that could be done by high winds should be considered. A reflector can increase available light to a collector surface by 40 to 50%. Also, don't underestimate the value of snow as a reflector. If you live in an area which experiences full winter snow cover, don't shovel it away from the south of the greenhouse. Leave it there and throw on clean snow when it gets dirty.

FIGURE 31

A third way to raise ambient air temperatures through controlled collection is to employ a venetian blind device. The Tucson Environmental Research Laboratory is experimenting with such a blind; it hangs between glazings in a window or in front of a Trombe or clear wall, creating convection currents of heated air within the unit (Figure 31). It is painted black. If the blind is partially or completely closed, its surface acts as a heat absorber, open it allows you to look through it. Because of its convective function, the Tucson researchers believe that on a cloudy day (diffuse radiation) the partially closed blind raises air temperatures more than direct transmission. (See Franta, page 103 Chapter VII).

Simple venetian blinds hung *beneath* the interior glazing of the clear roof of a greenhouse, painted black on one side and silver on the other, do several things. They raise the air temperatures before the air enters the home. This speeds the convection pattern and provides better circulation. The silver bottom would stop some radiation loss through the clear roof at night. In warm months, the silver side is turned up for an infinite number of shade settings.

Another way to raise ambient temperatures is to install a conventional flat plate collector *below* the level of the greenhouse, or in a low section of a clear wall. If there is adequate space for air passage between the two, a convective loop will establish itself (see Figure 96, page 132).

A more intricate passive way to raise temperatures is seen in the Pragtree Farm Unit (Figure 90, page 125).

INCREASING STORAGE

Increasing passive heat storage is a viable option for improving performance. Fifty-five gallon oil drums full of water are the most common form of "direct gain" storage. The basic problem with fifty-five gallon oil drums is that they are ugly (Figure 32). It would take a design genius to make them otherwise. Your family may not let you bring them into the greenhouse. For maximum efficiency, the drums should be painted flat black; this does not improve their appearance. They are cumbersome, usually bent and greasy. Some suggestions:

1) Buy new, shiny, unbent drums from the drum factory...highly unlikely, as the oil companies that own the factories choose to peddle them full or at $12 a whack empty.

2) Clean used barrels with Gunk Remover, treat them with a primer, then paint them a beautiful flat dark *earth* color instead of black.

3) Decorate the noncollecting areas of the drums to suit your fancy and taste.

4) Train plants around the sides and back of the barrels.

5) Incorporate shelving and planters on the top of the barrel installation.

FIGURE 32

However ugly they might be, the drums' effect on your greenhouse can be beautiful. Consider the fact that if direct sunlight raises the temperature of the water in a fifty-gallon drum thirty degrees, you will have stored about twelve thousand BTU's of heat energy. *Water stores three to four times as much energy per pound as an equivalent amount of rocks or masonry.*

The drums can be stacked near the back wall or wherever they will catch direct sun and not shade or be shaded. If they are stacked on end, stagger them slightly to allow space for filling and air circulation. If you have enough room, lay the barrels on their sides with the filling holes facing up. See that the drums *do not touch* the greenhouse walls or they will conduct heat to the walls that could otherwise be used to raise ambient air temperatures.

A variety of other water storage containers can be employed. They include water-filled beer cans (see Doug Davis, page 97), glass and plastic containers of all sizes, discarded gas tanks from cars and trucks, rubber and vinyl pillows, open metal vats and used galvanized steel tanks. The important thing is to check that no direct gain storage unintentionally shades storage behind it. For this reason various sized containers, lower to the south and higher to the north, are advisable.

Another direct-gain storage modification can be applied to inside walls that receive sunlight. This option is to add thermal mass to the walls by facing them with brick or dense stone. The massive facing should be of a dark color and nonreflective texture. It will store collected heat and give it up to the interior in the same way the water tanks do. The existing wall that you have covered will insulate against heat loss to the outside.

Be sure to "tie-in" the massive facing to the old wall with metal straps or lath to avoid separation. Memorable injuries could be sustained if the monolithic structure were to topple while you mulch your melons below. Rocks or bricks can be made into beautiful raised planting beds, doing the practical job of storing energy while improving thermal storage.

A similar modification can be made in the greenhouse floor. A thick concrete slab and/or stone paving added to the floor can act as a "heat sink" for storing thermal energy. The stone or concrete surface should also be dark and flat in color for maximum absorption. For best results, insulate against heat loss into the ground by putting a layer of pumice or styrofoam *under* the floor slab.

The soil in the planting beds can provide passive thermal storage. In the author's unit there is an estimated 4000 pounds of black mountain earth. Again, an insulator like dry pumice, gravel or styrofoam (with holes for drainage) should be under the storage.

DECREASING LOSSES

Let's take a look at the second option I mentioned: decreasing heat losses/increasing insulation. These two areas of improvement go hand in hand. For starters, if you have a wind chill factor *inside* the unit, that's a problem. Places where those little zephyrs whistle through the structure should be plugged up. They usually occur around door and window frames, roof and wall junctions, plastic to wood joints...let's face it—anywhere you have put two pieces of the greenhouse together. These "infiltration losses" are the most common and the easiest type to correct. Simply grab most any pliable material handy (hopefully with some insulative value) and cram it into the crack or hole with a screwdriver or bent coat hanger. Leftover fiberglas insulation pulled cotton-candy style off its backing is recommended.

Some air leaks, such as those that occur where a vent or door meets the jamb, will require conventional weather stripping. A caulking gun full of sealant will eliminate most others. Squeeze an even "bead" of caulking material along all joints that pose a problem.

Convective and conductive heat losses are those caused by air movement *within* the walls, roof and clear-wall areas of the greenhouse. Reducing these is more difficult than plugging leaks. First check whether you have cold spots on solid walls or in the insulated roof. Go into the greenhouse after a cold night and feel around on the walls. Your hands will tell you if you have an isolated area of high heat loss. It will be noticeably colder than the rest of the walls there. If you find the culprit, add more insulation. In a frame wall, put it on the interior side or in the wall. On a massive wall, put it outside.

In a well-sealed greenhouse, the greatest losses will occur through the clear wall and clear roof areas. Traditional (all clear, exposed) greenhouses rank among the most energy *in*efficient structures ever built. That is why solar greenhouse designs use clear surfaces only where absolutely necessary for heat collection and plant growth. Still, energy will be lost at an amazing rate through the skin when the sun goes down.

Multi-layered (more than two) glazings add some insulation to the skin but also decrease the transmission of light by about 10% for each additional layer. Recent studies have found that a triple glazing is justified in areas that have full winter snow cover. The added reflection of the snow makes up for the loss in transmission. Several universities and chemical companies are experimenting with a poly-film application that is opaque to the infra-red on the inside. That would keep much more heat in polyethylene greenhouses.

One time-tested device that can insulate your clear walls is a curtain or drapery. Pulled closed in the evenings or on cloudy days, the curtain should seal as tightly as possible to be effective. Ideally the curtain should have a reflective surface (such as aluminized mylar or tin foil) *facing the air space* in the interior. A disadvantage to the curtain is that you have to be present to operate it (unless, of course you install a motor with a heat or light-activated switch).

A more simple approach is to use rigid foam panels to fill the clear areas at night. Clips and magnets have been used to attach the rigid panels (see Zomeworks, page 130, Chapter VII). The obvious disadvantage is that the panels must be removed and stored during collection periods. You might devise a mounting system for the stored panels in which they double

FIGURE 33

as interior reflecting surfaces to combat phototropic plant growth.

Another type of insulating curtain is made from rigid styrofoam or polyurethane panels sandwiched together with cloth on alternate edges (Figure 33, page 59). This oriental-style curtain is stored at the top of the clear wall and lowered via a cord or wire that runs through the panel centers from top to bottom.

In any vertical curtain arrangement, the fit must be tight when closed or the system may defeat itself due to the "chimney effect." Thermal action can set up convection currents between a loose curtain and the clear wall, causing heat loss greater than were the curtain not there. A horizontal curtain similar to the one discussed in Appendix C, page 155, but manually operated, can be pulled across the interior of your greenhouse. It would keep more heat down near the thermal storage.

Insulating materials on the outside of the structure offer another solution to the heat loss problem. The one shown in Figure 30, page 55, that doubles as a reflector is an example.

Any blanketing system would help maintain higher nighttime temperatures in very cold weather. This winter I will be using exterior rolldown blankets made for covering cement. They are 1 - 3" thick with a flexible insulation interior covered with waterproof black polyethylene. Cement companies sell or rent them.

ACTIVE SYSTEMS

Passive systems all demand your presence for their operation. An alternative, an automatic insulating wall, has been developed by Zomeworks in Albuquerque, New Mexico. The design is *active* in that it uses electrical blowers to fill a double-glass clear wall with small styrofoam beads when temperatures drop. The "Beadwall"® acts as a solid insulated barrier when filled, and a transmitting clear wall when the beads are automatically vacuumed out.

With the Beadwall® we get into active system options for improving greenhouse performance. An active system involves collecting, then moving heat energy (via water or air) with a pump or blower. It is then transferred to a storage medium (rock and/or water) for immediate or later use. Heat of Fusion Storage, using salt hydrates as the storage medium, is a third alternative, but as we know of no present applications in greenhouses, we'll omit it here.

One relatively simple active hot-air system that has been used in solar greenhouses is shown in Figure 34. Hot air is tapped from a high area in the unit and then ducted beneath floor level to a rock bed where it is stored. The fan can be activated by a thermostat that cuts in when the ambient air temperature reaches seventy degrees. At night and on cloudy days, the rocks radiate heat into the unit. Alternatively, a second fan can be used to direct the heated air through the rock bed into the greenhouse. For working examples of this system, see Chapter VII, pages 100 - 101.

A successful variation of this system features *exposed* rock storage through which hot air from the greenhouse ceiling is pumped (see Figure 35). Exposed rocks are contained by wire mesh and serve as table platforms (see Figure 36). The rocks store heat from direct sunlight as well as from the heated air fanned through ducts from the ceiling (see Jemez House, Chapter VII, page 93). Doug Balcomb (Chapter VII, page 111) has chosen to build his rock bin in an insulated box inside the greenhouse. The beauty of this system is that it can be added later if additional storage is needed.

Another active system option is to use an *enclosed* solar heater (air or water) mounted on the greenhouse roof or inside the unit. If it is a water system, liquid will be pumped through the collector into a large storage tank (possibly underground, surrounded by rocks and insulated). When the sun goes down, heated water bypasses the cool collector, making a closed circuit from the hot storage tank through baseboard radiators in the greenhouse or a conventional heat-transfer blower.

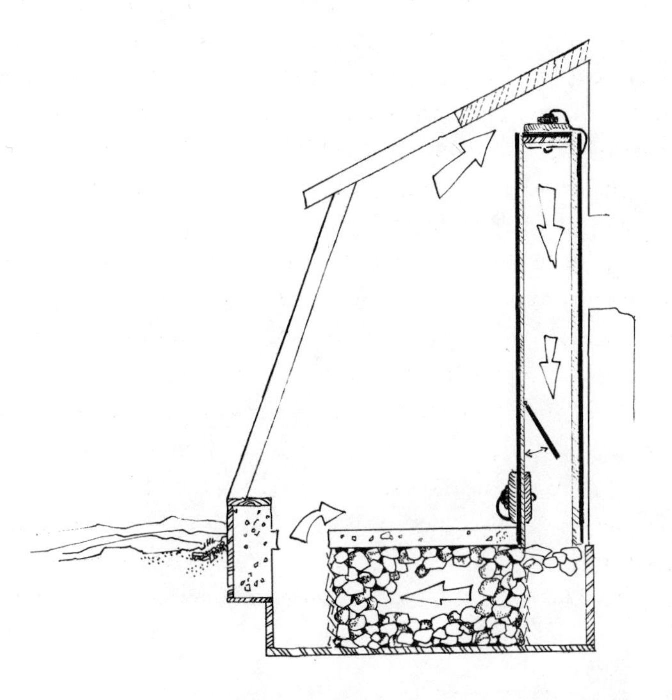

FIGURE 34

FIGURE 35

Disadvantages of this type of system are the high initial cost and the technical skill required for its success. Another drawback to a pressurized and circulating hot-water system is that if the electrical power to pump it falis, the water stagnates (stops moving) and superheats, blowing out the entire system—or at least the pressure relief valves. This can be discouraging.

Components such as collector boxes, pumps, blowers and so forth are commercially available, but the final arrangement, or system design, is left largely up to the individual. I suggest that if you plan to incorporate a complex active system into your greenhouse, *buy* the more important components rather than building them yourself. Playing the solar inventor is an exciting pastime, but it can be very expensive and give birth to many a white elephant. My first solar collector box clung precariously to the roof, rusting and rotting in the weather. Its vacuum cleaner blower sounded like a freight train howling through the house. In short, it was a $150 mistake. If you do feel qualified to construct a collector, at least follow plans for units that are proven effective.

FIGURE 36

A proven and inexpensive water heating system is the "batch" tank or "bread box." The unit is used as a pre-heater for your conventional hot water or directly in small applications. It consists of a tank (40 - 60 gallons) enclosed in a reflective box and mounted in the apex of the greenhouse. It runs off the home line pressure and, because of the large volume, doesn't run the risk of overheating. To prevent the danger of freezing out, leave large holes in the bottom (greenhouse facing) cover or make a moveable reflecting/insulating exterior cover. Still scared of freezing? Bypass and drain the system for the coldest time period.

A forty gallon tank like this will raise water temperatures to the gas heater about 40° Zomeworks (Chapter VII, page 130), sells plans for bread box heaters.

Complicated high temperature hot water systems cannot presently be economically justified in greenhouses. Avoid them. Use air or low temperature water as the heat exchange medium, if an active system is what you need. Refer to the rock storage systems in this chapter or in Chapter VII. You will be surprised at the performance of the passive units and the simple improvements we've described.

CONVENTIONAL HEATING

The third alternative for improving greenhouse performance is to employ conventional supplementary heating. The solar purist will object violently, but there is a great deal of validity in this option. If you have built your solar greenhouse properly, the amount of conventional fuel needed to improve performance should be minimal. For the greenhouse operator in extreme climates who shuts down the unit for a month or so in the winter, the time period when supplementary heating may be needed will be brief.

A small gas or electric heater that cuts in at 35 or 40 degrees can provide an early beginning for the late winter/spring crops and a healthy harvest in the fall/early winter period. The heater shouldn't be necessary that often, and on an economic level could be justified by the yield of produce and flowers. On a purely emotional level, a small heater can eliminate your anxiety over those tender young seedlings that have so bravely sprouted in late January.

In my greenhouse I rely on a wood-fired pot belly stove in an adjoining room for emergencies. I stoke it up on cold, clear nights following snowy or cloudy days. As the effort and wood consumption isn't great, I feel justified in using this option.

COOLING THE GREENHOUSE

Thus far we have been concerned exclusively with keeping the greenhouse warm. Another common difficulty arises from the opposite problem: maintaining tolerably cool temperature levels in summertime. Again, if you have built in the correct amount of solid roof area for shading, provided proper ventilation and thermal mass, this should be controllable. If your unit still overheats, however, screen shading is one way of correcting the situation. Greenhouse suppliers market a number of plastic sun screens that can cut down the incoming light anywhere from 15 to 85 percent. Bamboo curtains are another possibility; they are usually more expensive but to many, worth the extra cost (see Figure 37). Bamboo-like plastic curtains are effective and cheap. Get white or a light color.

A traditional method of cooling greenhouses is to whitewash the outside of the clear surfaces in the summer. This is time consuming, relatively unattractive, and could have adverse effects on fiberglass-acrylic products.

Another type of passive cooling involves placing a small pan of water in front of the lower vent. From the top of the vent hang strips of burlap, allowing them to dangle down into the water. Water will seep up the burlap (the wick effect). The natural convection of the heated air exiting from the high vent will pull evaporatively cooled air through the greenhouse.

An exhaust fan is inexpensive, uses little power and can be mounted in the upper vent

FIGURE 37

of the greenhouse. We have never had to use it in our designs but would as a last resort.

Solutions that we have used effectively for cooling include taping large sheets of butcher paper or newsprint to the inside plastic, and/or growing walls of shading plants (inside or out) to block incoming light (see Chapter VI, page 76).

A mist spraying, or watering of the floor, rocks and earth beds around noon on hot summer days will lower temperatures 8 to 12°. This has been very effective for us.

Cooling vents can be opened and closed by heat activated pistons. These are totally passive mechanisms and only moderately expensive.

If you have made the mistake of buying a poorly designed commercial unit that overheats, try large shading panels of masonite or sheets of opaque plastic on the roof.

In combatting overheating problems, remember the general rule that low-angle winter sunlight is to be allowed into the unit, while high-angle summer rays are to be partially blocked. A common misconception is that all plants must have direct sunlight for healthy growth. This is simply not the case. Also, many books will tell you that the greenhouse must stay under 85° and not get colder than 55°. That may be true for orchids, but most vegetables are hardier. They can thrive in much higher and lower temperatures than that. Careful planning and placement of your crops, and the use of reflected light, will get tangible results for your enjoyment and nourishment.

CHAPTER VI The Greenhouse Garden

One of the most innovative aspects of the Solar Sustenance Project greenhouses is that they are geared mainly to the production of vegetables—food crops—rather than flowers and house plants. Food production (and surplus solar heat) transforms the greenhouse from a luxury item to a functional addition to the home. The Solar Sustenance Project units have produced fine crops; their size and variety is dependent, of course, on the amount of time and care the owner can offer.

Gardening is a subject that fills volumes (as you well know, if you've browsed through that section of a bookstore lately). Check our bibliography for books that cover the whole realm of gardening. In this chapter we are going to have to limit ourselves to plants that *we have grown in our greenhouses* and can tell you about with honesty and confidence. However, we are always trying new ones and different planting arrangements, and we hope you will do likewise.

We should point out that the greenhouse gardener has some advantages and problems that the outdoor gardener does not have. Planting space is much more limited, for one thing. Also, an attached greenhouse is an integral part of your living environment; your home shares in the problems of the unit (insects), as well as deriving benefits from it.

The greenhouse gardener does have one great advantage over the outdoor gardener: you are not as dependent on the weather. Low temperatures in the coldest part of the winter are the only exception. The ravages of wind, hail, drought and frost have no effect on greenhouse plants, whereas outdoors they can and do destroy many crops. In the high mountain valley where the Fishers live, midsummer hailstorms are a constant threat. The farmers there say that they don't call a crop successful until the food is on the table. As a greenhouse gardener, you enjoy both a greatly extended *and* a much more predictable growing season, due to your control over the climate in your unit.

GREENHOUSE LAYOUT

The size, shape and number of planting areas included in your greenhouse will be determined by the configuration of the unit and your own judgement (practical and aesthetic). However, since space is limited, plan the layout of planting areas with the utmost care. For maximum productivity, leave very narrow walkways and put the rest of the floor area into beds and tables.

Vertical space should also be used to the greatest extent possible and you'll find that there's quite a bit of it. Shelves lining solid walls and hanging planters are two methods I have used. I've also installed moveable shelves on brackets along the clear walls to hold

FIGURE 38

trays of seedlings. By June these shelves can be taken down and stored (Figure 38). Temporary tables can also be erected in beds for seedlings (Figure 39). Another good way of using vertical space was devised by the University of Arizona Environmental Research Laboratory. It involves a series of hanging pots, attached one above the other (Figure 40). This ingenious idea really makes the most of greenhouse growth/space potential.

Some of the space in your unit will have to be given up to equipment. If you already have a tool shed, then you won't need too much storage area in the greenhouse. But, empty trays and pots, cartons of seedlings, small hand tools, containers of fertilizer and other basic necessities of greenhouse living do accumulate quickly. When planning the interior layout, leave space for equipment and a comfortable amount of standing room.

FIGURE 39

SOIL

Gardeners used to "taste" their soil before planting. If it tasted sour or bitter, it was not good for raising plants, but if it tasted sweet, a high yield could be expected. The soil that tasted sour was too acid; soil that tasted bitter was too alkaline. A good balance between the acidity and alkalinity of soil is important for a good harvest. If you prefer not to, it is not necessary to taste the soil. Instead, send it to a county agent or buy a do-it-yourself soil testing kit.

A healthy plant needs Nitrogen (N), Phosphorous (P) and Potash (K) from the soil; trace mineral elements are also necessary. Nitrogen is needed for growth, particularly for green leaves. Yellow leaves and poor growth denote a lack of nitrogen in the soil. Organic matter decaying to humus releases nitrogen slowly, but is a steady source. Cottonseed meal is another excellent source of nitrogen. Phosphorous is needed for roots, fruit development and resistance to disease. It comes from organic decaying matter and from ground phosphate rock. Potash is needed for the cell structure of the plant. It comes from organic matter, from wood ashes and from green sand. Trace mineral elements come from a variety of components in organic matter. Most plants like a neutral or slightly acid soil.

FIGURE 40

Your soil should be 12 to 16 inches deep. Good drainage is a necessity. There should be 6" to 8" of pumice, sand, vermiculite or pea gravel under all of the soil in your beds. We have not tried it in our greenhouses, but from the reading we have done, we feel that the Biodynamic French Intensive method of gardening will probably adapt well to the greenhouse but will need deeper beds. Contrary to traditional greenhouse techniques, no one in the Solar Sustenance Project (about 25 producing units) has ever sterilized their soil. You do not sterilize your outdoor garden soil, do you? Besides, how can you get that much dirt into the oven?

Greenhouse soil should not be *as dense* as regular garden soil. It is lightened by adding other materials. Two good soil mixtures are:

¼ rich organic topsoil
¼ coarse sand
¼ peat
¼ vermiculite or perlite

If you live near the mountains, try this one:

1/3 black mountain earth (found under the pines)
1/3 arroyo sand (river bank)
1/3 pumice, perlite or vermiculite.

Soil can be kept and used for a long time in a greenhouse, but it must be regularly enriched with compost and organic matter. Rotating plants and using nitrogen fixers, such as peas, in various locations helps keep the soil viable. Mulching cuts down on water requirements and does add some nutrients to the soil. (See later sections on crop rotation and fertilization.)

The soil in the greenhouse must drain well. Plants do not like sitting in a puddle of water. If your soil is on the heavy side, add a good portion of vermiculite and organic matter. Air is another vital ingredient in the composition of healthy, fertile soil and in producing well-rooted plants. Along with using organic matter, another way to aerate and allow water to penetrate the soil is by providing a good supply of earthworms. They fertilize soil by means of casts which are rich in nitrogen, potash and phosphate. Earthworms also help balance the composition of the soil; if your soil has a goodly number of earthworms in it, you do not need to add nutrients or worry as much about insects. There are over two thousand kinds of earthworms but the two which are of interest in the greenhouse are the soil ingesting blue and the manure and compost ingesting red. Both make humus.

HYDROPONICS VS. SOIL

Hydroponics is a method of highly controlled agriculture whereby plants are grown in a "non-soil" medium such as sand or vermiculite. A balanced diet of nutrients is fed directly to the root system.

There is a long standing battle raging between hydroponic and soil aficionados. Hydroponic gardeners claim as high as 300% increases in yield; soil users retort, "Hydroponic plants are chemical junkies and besides they don't taste good." I have limited experience with hydroponics but have found the following:

1) Hydroponic plants are more dependent on continuous care. A soil plant in a deep bed can survive several days in midsummer without watering. A hydroponic plant will die if its feeding is missed for a day. For this reason, most hydroponic operations call for an electric timer, feed batch tank, tubes to plants or containers for feeding and drainage. In a small, home operation, these items are not particularly expensive or energy consuming.

2) I have seen plant growth and densities in soil that are not surpassed by any hydroponics techniques (for example, the Tyson's greenhouse, Chapter VII, page 89, and photo on next page). Given, this is a small operation which receives a maximum of tender love and care. In a larger commercial unit increased yields by hydroponics are documented.

3) The soil in beds and on tables has the ability to store heat. Some hydroponic media such as pea gravel can also thermally charge but other media such as vermiculite does not hold much thermal storage capability.

4) For the complete novice I recommend starting with soil. The greenhouse is a new experience by itself. The added element of hydroponics might complicate things. After you have had some experience with the greenhouse, run a hydroponic experiment and then decide which side of the battle to join.

We have included some good hydroponic books in the reference list.

FIGURE 41

FERTILIZERS

Every time a crop is harvested, the soil should be fertilized. Remember: you're asking a great deal more of the greenhouse earth than the garden earth in terms of length of the growing season and density of the crop. Because of the relatively limited space, most greenhouse owners plant intensively. Many of them, including the Fishers and Yandas, deliberately crowd their crops in order to squeeze every last tomato, bean and pea from the unit that they can. This type of intensive planting does work, if the soil is rich enough. Regular fertilization is the primary way to enrich the soil.

There are many different types of fertilizers. A local nursery can recommend a variety of basic dry and liquid commercial brands. If you apply any of them every time you're ready to plant a new crop, the soil should stay in good condition. Your plants are the best indicator of insufficient fertilization.

Organic fertilizers, such as manure and compost, work beautifully. We have found dry cow manure to be very effective. Keep in mind, though, that every time you bring foreign matter into the unit, you run the risk of bringing harmful insects with it.

Manure and compost should also be added to the soil while the plants are growing and producing. Frances Tyson makes what she calls "horse manure tea" by soaking fresh dung in a pail of water. Frances periodically pours this hearty brew on her soil; the health and beauty of her plants (see Figure 41) testifies to its value as a fertilizer.

Many people believe that organic fertilizer is more effective overall than chemical fertilizer (we're among this group). We have used fish emulsion very successfully. I like the stuff; it makes my whole house smell like Provincetown, Massachusetts, for a couple of days. But what you use depends to a great extent on what is available. The most important thing is to fertilize regularly, lavishly and deeply. When fertilizing, turn over the soil with a shovel or pitchfork. Dig down to the bottom and really move that earth around. You will bury the fertilizer deep into the soil, the best place for it. It will feed the whole root system of your plants and also encourage the roots to grow deeper, thus creating a firmly and well-rooted crop.

FIGURE 42

Mulching is another good way to enrich the soil and keep it evenly moist. Mulch with organic substances that break down readily into humus. Grass clippings, leaves, wood shavings, vegetables and other leftovers—anything you would compost—will work very well in the greenhouse. Frances Tyson has successfully turned her beds into compost piles, using nearly all her organic garbage as mulch and fertilizer. She keeps a small jug by her sink and every day or so turns her "garbage" into her greenhouse soil.

Get an image in your mind of the soil as a living element. Nourish and care for it just as you do your plants. The soil essentially makes the difference between a sterile and a fertile greenhouse (Figure 42).

PLANTING LAYOUT

Before planting, give careful thought to the light requirements, warmth needs, sizes and shapes of various crops. Plants in the front beds will get more direct sun. (See Figures 43 and 44.) I have found that all the crops I've planted grow vigorously in the front beds, so the real challenge is finding plants that will grow on the north side.

Profile: SPRING–SUMMER–FALL

A. Shading plants
B. Fruiters. Tomatoes, cucumbers trained up twine. Trim foliage, squash, melons.
C. Seedlings, herbs, fruiters. Hydroponic table.
D. Low light, coolest greens. In late summer new fruiters can go here. Climbers, flowers.
E. Flowers, shade lovers.
F. Berries, shade lovers.

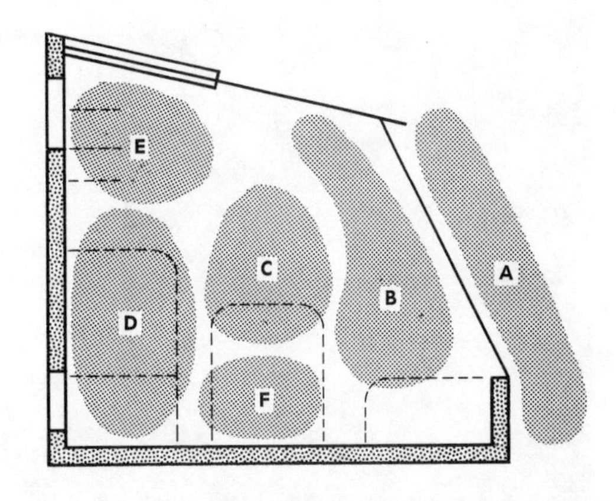

FIGURE 43

Peas seem to do very well in back beds; broccoli and herbs do all right, also. I have never given my pole beans a choice; they are automatically planted in the most unobtrusive bed, in the back eastern corner, where they still receive enough sun but hardly shade the other crops. Generations of beans have been very content in that bed.

Profile: WINTER

A. Lightest, coldest. Leafy greens, radishes, peas, broccoli, roots, tubers. Carry over fruiters.
B. Light, cool. Herbs, greens, flowers. Transplant seedlings.
C. Light, warm, hanging pots, flowers, herbs.
D. Light, warm. Winter tomatoes, peppers. Climbers, beans, houseplants.
E. Low light, warmest. Start seeds, sprouts. (On shelves, bread will rise well here.)
F. Shady, cool. Berries

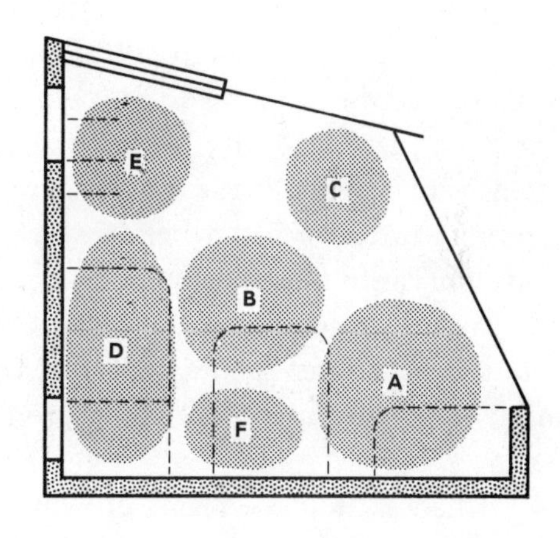

FIGURE 44

You will probably want to experiment with a variety of planting layouts. Just remember to give extra attention to crops in the rear. Provide them with as much light as possible through the use of reflective surfaces as described in Chapter V.

In the overshaded greenhouse, plants tend to become phototropic; i.e., they grow towards the sun in the direction of the clear walls. Extreme phototropism can affect the plants' vigor and ability to produce; again, you can combat this problem with the use of reflective

FIGURE 45

devices along the solid walls. Also, potted plants and seedlings for transplanting may need to be turned, just as they do on a windowsill, in order to compensate for phototropism.

As you get ready to sow your seeds, keep in mind that most greenhouse owners plant intensively and that it can work quite well (see Figure 45). The number of plants that you can cram into one bed depends a great deal on the fertility of the soil. Under good conditions you can cut the recommended spacing by ¾ths for many crops and get a fine harvest.

Intensive planting may affect the appearance of some crops. My chili peppers (spaced about two to three inches apart) were spindly plants that had to mutually support each other to stand upright. But they produced an excellent harvest, each one laden with peppers.

I had six tomato plants growing in a 5 x 3' bed. By July they were massive and produced all the tomatoes we could eat with plenty for canning. The plants were overflowing the borders of the bed and advancing down the walkways by October; still they couldn't seem to stop producing. By the end of December they were completely out of control, threatening to take over the adjoining room in the house. I harvested grocery bags full of tomatoes just before my winter freezeout.

On the basis of these and other experiences, I'd say that intensive planting can be very successful in the greenhouse. Just keep in mind the nourishment your soil will need in return.

THE VEGETABLE PLANTING CYCLE

Don't grow any vegetable that you don't enjoy eating. It's no major accomplishment to grow radishes or other hardy cold-weather crops in your solar greenhouse at any time during the year. But if you dislike radishes, then it's a waste of time and soil nutrients. Raise the vegetables that you and your family eat and your success will be measured not only in terms of what you *can* grow, but what you use.

Having said this, we encourage you to plant *any* vegetables or flowers that you especially like. If you take proper care of the plants, start them at the right time of year, ward off insects and follow the other basic procedures we are suggesting, there is no reason why you can't grow almost anything you want to. We urge you to experiment, particularly if you enjoy gardening enough to take a few risks in return for possibly spectacular results. Imagine how you would like your greenhouse to look and them simply bring that vision into reality. It can be done.

Thus far we have talked mainly about vegetables, but don't feel compelled to use every square inch of the greenhouse for edibles. Flowers add color, fragrance and beauty to your unit year round. If the greenhouse space is an attractive one, you'll find that you want to spend time in it and maintaining the plants will be a pleasure. The plants will enjoy and benefit from your company. This makes a crucial difference in the success of a greenhouse.

Greenhouse seasons are different than those outdoors. Summer comes early and lingers on. The planting cycle in the greenhouse is based on this extended summer. The Fishers have divided the cycle into three time periods: late winter/spring, summer, and fall/early winter. The actual months these periods cover run from about mid-February through May, June through mid-September, and mid-September through mid-December. Some crops will be harvested within one period, others will extend over two cycles and still others, notably tomatoes, will continue to bear through all three cycles.

The following planting cycles are based on first-hand experiences in the Fisher and Yanda greenhouses. We are including both as examples of different modes of operation in two different climates. It might be noted that *neither* of these greenhouses have any water storage or clear wall insulating covers.

The Fisher Greenhouse

The Fisher greenhouse is located in a mountain valley in northern New Mexico, altitude 8,000 feet, 36 degrees north latitude. El Valle experiences a high number of cloudy days and the last killing frost is usually in early-to-mid-June. The unit is attached to an adobe house. The design of the greenhouse is quite similar to that described in Chapter IV, but it has a vertical front face rather than a tilted one. The Fishers have chosen to shut the unit down for two months each year (winter freezeout) rather than add supplemental heat or other improvements at this time. The planting cycle in the Fisher greenhouse is as follows:

Late winter/spring. In mid-February I start tomatoes in peat pots or Jiffy pots. On very cold nights they can be moved into the house. Peas, lettuce, onions, broccoli and radishes (hardy cold weather crops) can be planted directly in the greenhouse beds. I recommend succession planting for the lettuce.

FIGURE 46

Toward the end of March these late winter/spring crops are going strong. At that time I transplant the tomatoes and peppers into the beds. I put the peppers in with the lettuce as it will be harvested by the time the peppers really start to fill the bed. The tomatoes, which are heavy feeders, go in a bed by themselves since they grow more rapidly and will continue to produce through all three seasons.

At this time I also plant pole beans, a very successful crop in my greenhouse (see Figure 46). They put most of their energy into vertical growth and thus do not take up much space in relation to the amount of beans they produce. You do have to provide a latticework of stakes for them to climb and it should extend to the ceiling, as the beans will easily reach that height.

Any additional summer crops I want to plant, such as cucumbers, cantaloupe, pumpkins and annual herbs, also go in by the end of March. I've tried corn, too, which was flourishing in the greenhouse and producing ears in early July. The ears didn't seem to be developing, though, and when I pulled back the husks to check them, I found hundreds of aphids devouring them. If it hadn't been for these insect pests, I see no reason why the corn plants wouldn't have done very well in the greenhouse. Corn is, however, a crop that takes up quite a bit of space relative to the amount of food it produces.

By mid-April my greenhouse is in full swing. The peas, another fine crop for the unit, are a mass of blossoms and I'm getting peas daily. The late winter/spring cycle is perfect for peas which don't do well later in the midsummer heat. Snow or sugar peas, the variety I usually grow, seem to do especially well in a cool greenhouse.

Lettuce is in its prime at this point, also. I harvest the outer leaves and let the interior keep growing, except for those heads that need to be thinned out of the bed. Lettuce is one of the hardiest crops in my unit and has done well through the coldest nights of the year.

Radishes are over by now and I can replant if I want to but usually I give their space to a summer crop. Onion greens have, of course, been edible all along but the bulbs themselves are slow to grow (and I've never had much luck with them). Now I generally pull them up when they are still green and young and enjoy them in salads. Broccoli is large and sturdy in mid-April, getting ready to head.

Mid-April is also the time when I start vegetables in peat (or Jiffy) pots for transplanting to the garden. They will have six to eight weeks to germinate and grow in their pots before going outdoors, which is just about right. If you start them much earlier, the plants will outgrow their pots, get badly rootbound and may be damaged before it's safe to transplant them. This is more of a problem with cucumbers and melons, which cannot tolerate root damage, than with plants such as tomatoes and peppers, which are hardly affected by it.

Summer. By the end of May the lettuce and peas are harvested. The pole beans have reached the ceiling and are still blooming. Tomatoes are covered with flowers and have green fruit on them. Cucumbers and melons are beginning to produce, too. The pepper plants are growing rapidly and the broccoli is still heading. The herbs can be used daily, especially parsley, which is a prolific producer. The crops that were started in Jiffy pots can be hardened off (exposed to harsher outdoor weather) and transplanted to the garden.

In my experience, some vegetables that were planted directly in the beds can be moved outdoors even though they are quite large by this time. I have transplanted tomatoes, peppers and broccoli, blossoming and fruiting, with no ill effects on their productivity. My broccoli actually benefitted from the move as the garden was cooler than the greenhouse.

Fall/early winter. In early September parts of my greenhouse take on an autumn look. The bean and pepper plants are fading; the cucumbers, cantaloupes and pumpkins are nearing their end, although some of the fruit is still ripening. I remove all the summer crops except the tomatoes and herbs. Then I fertilize and get ready to plant again.

Lettuce, peas, broccoli, and a second planting of pole beans are my fall/early winter crops. They will thrive during this period although the beans do not produce quite as well in December. The unit's summerlike warmth in the fall promotes fast germination and the cool temperatures later on are perfect for these vegetables (see Figure 47). Through succession planting I *can* keep peas, lettuce and broccoli going all winter, but I prefer to give the greenhouse (and myself) a time of rest.

FIGURE 47

The Yanda Greenhouse

The Yanda greenhouse is located 15 miles north of Santa Fe, altitude 6400 ft., 36° north latitude. The last killing frost in this area is usually late May. The front face slope of this greenhouse is 60°; it is run year round. The following describes the Yanda's planting and maintenance schedule and the different vegetables they have planted and worked with.

Nambe greenhouse Planting and Maintenance Schedule

Mid-January: Plant cold weather crops—beets, broccoli, peas, radishes, spinach, lettuce, any herbs, garlic, nasturtiums, marigolds. In the house start tomatoes and chili peppers for transplanting into the greenhouse in several weeks.

Mid-February: Transplant starts into the greenhouse. Start any permanent plant except very warm weather crops such as cucumbers and melons.

Mid-March: This is a deceptive time of year. Everything in the unit is growing like mad and you are already harvesting. You are tempted to start outdoor garden seedlings. Don't do it yet. They will become rootbound or get too big for transplanting. Start your melons and cucumbers if you plan to leave them in the greenhouse.

Mid-April: Now start seeds for garden transplant.

Mid-May: *Harden off* garden transplants by exposing them to harsher outdoor weather. Do this for about two weeks. Start some tall or climbing plants for summer shading. Sunflowers (see Figure 48), morning glories, hollyhocks, etc., can be used outdoors on the south face; pole beans have been used inside.

FIGURE 48

Summer: This is a time of heavy production in the greenhouse. Keep the greenhouse clean, well ventilated and the insects under control. (Note: the greenhouse should always be kept clean and weeded.) Train climbing vegetables up on strings, keep vegetation trimmed. Help pollinate certain crops. Tap tomato blossoms lightly at the base of each cluster; use a small paintbrush, dabbing the male, then female flowers of the melons and of the cucumbers. Enjoy your produce.

Mid-August: Start vegetables for fall and winter harvesting in the greenhouse. Such vegetables as broccoli, peas and beans should be fully grown before the really cold weather sets in.

September: Some outdoor garden plants can be transplanted back into the greenhouse. Check for insects and disease first.

Fall-January: Check the greenhouse for any repairs that need to be done. Continue to pollinate and trim back foliage on fruiting vegetables in front beds.

Mid-December to Mid-January: This is the coldest time of the year and because of short daylight hours all plants show a slow growth rate. You may find this a good time to really clean out the greenhouse; wash all containers and boxes, replenish the soil. This is an excellent time to get rid of any lingering pests by freezing them to death. Simply open up the greenhouse at night for a week to ten days. Close it during the day and turn the soil several times. Like us, you may find this the most rewarding time of the year to have a garden. Hardy and cold weather plants will grow during this period without added heat.

The planting chart (Appendix B) will give you a full listing of varieties, time of planting and locations we've used in our greenhouse.

FLOWERS IN THE GREENHOUSE

Houseplants: In my experience, all houseplants thrive in the solar greenhouse. Many people give their plants an invigorating outdoor vacation during the summer. A vacation in the greenhouse is even better. The color, size and density of foliage improves tremendously, and the plants look as good or better than when they were new. Therefore, I recommend that any plant suffering from the windowsill, coffee table or bookcase blues be transferred into the greenhouse for a couple of months or more.

Light and humidity in the greenhouse are two important factors in reviving plants. Be sure, though, to place plants that are susceptible to burning in a location away from direct sunlight, just as you would in your house.

Hardy houseplants can stay in the greenhouse year round, unless you are planning a solid winter freezeout. Geraniums are one of the best examples of this. At about 22 degrees, mine died down to the roots, then made a magnificent comeback in the spring. If your greenhouse temperatures never drop that low, cut back the geraniums in midwinter anyway. By April they will be beautiful again.

My strawberry begonias have also proven to be very frost-resistant houseplants. In fact, they seem to thrive at cooler temperatures. Wandering jews have been remarkably hardy in my unit as well, though again they die back when the temperatures are in the twenties. The color and vigor of the new leaves in spring is incredible.

More tender houseplants, such as the coleus, benefit greatly from the greenhouse environment. But they must be brought back into the house during the coldest months unless your unit never freezes. In this category, I have found that fuchsias, swedish ivy and spider plants are especially content in the greenhouse. I treat them exactly as I do in the house, in terms of watering, trimming and feeding.

Garden plants. All the flowering plants I normally have in my garden grow beautifully in the greenhouse. Choosing those to include in the unit involves considerations of space, size and general planting layout. I prefer smaller varieties that add color but don't require much space. My favorite in this category and one of the easiest is probably the petunia. They come in a wide range of colors and bloom profusely. If petunias get spindly, cut them back, and you'll see them return in healthy vigorous condition. Petunias are considered tender annuals, but in my greenhouse they have become perennials (hardy to about 22 degrees). The same applies to marigolds and pansies. Pansies, like petunias, need to be cut back periodically. All these types of flowers bloom continuously.

I have a dark moist corner in my unit where nothing seems to grow—except ivy. I bought a variety that is recommended for north sides of walls (Engleman's Ivy) and my problem corner has been transformed into a haven of green.

The gladiolus bulbs I planted along the border of the front beds have grown very well. In this region, gladiolus are not reliably hardy and bulbs planted in the garden need to be dug up and brought in during the winter. In my greenhouse, however, they remain in the ground year round; they die back to the bulbs in late fall, proliferate and return with increased beauty in the spring.

Transplanting. The greenhouse provides prime conditions for starting flowers to transplant into the garden. You can get a good headstart on the outdoor growing season and thus get much more enjoyment from your flowers during the summer.

In early April I start perennials such as poppies, daisies and columbine in the greenhouse; they can be transplanted sooner than annuals and will usually bloom the first year with this kind of headstart. I plant annuals, including marigolds, zinnias, morning glories and sweet peas in mid-April.

One of the advantages to starting flowers this way is that April temperatures in the greenhouse are warmer and more consistent than outdoor temperatures here in early June. Seeds therefore germinate faster and I can keep the soil evenly moist during the crucial time of germination.

I can also set the plants out, already several inches high and hardy, in any configuration I like. They are properly spaced and I know they will remain that way. I have never been able to achieve this kind of symmetry (without a lot of effort and frustration) when sowing seeds directly in garden beds. Starting the summer garden in the greenhouse is thus a great joy for me and has yielded excellent results.

Many flowers can also be transplanted into the greenhouse in fall. If you are careful to get all or most of the roots, they will continue blooming all winter. Petunias, pansies and marigolds are the three I've transplanted successfully.

Generally it doesn't make much sense to dig up hardy perennials that are well established in the garden, but you should follow your own preferences in deciding what flowers to transplant. When you do get ready to bring them in, dig up the roots with plenty of soil around them. This *lessens* the shock of transplant, decreases the danger of damaging the roots and gives the plants the illusion that nothing unusual has happened to them. By the time their roots have grown beyond that ball of soil, they should be firmly grounded and ready to cope with any differences in the growing conditions.

MAINTENANCE

Watering. The amount of watering that greenhouse plants need varies according to the season and climate. In summer and over prolonged periods of clear days (any time of the year) they will require more water than on cloudy days and during most of the winter. Determine the amount of watering you will need based on personal observation and experience. The best way is to stick your index finger down into the soil. The soil should be moist, not slushy.

DO NOT OVERWATER! A greenhouse uses much less water than an outdoor garden. Humidity should not be a problem in the greenhouse if you run it properly. Overwatering can rot the roots and create other problems. Use rainwater whenever possible; it is good for the plants.

In warm weather the best time to water the greenhouse is about noon. In summer evaporating water cools off the unit and keeps the temperatures comfortable during the afternoon. In winter a morning watering with lukewarm water will help the plants after a cold night. During the cold months it is best not to water on cloudy days.

The equipment you use can be as simple as a large watering can. In the summertime you can water the greenhouse with a hose from out of doors. Ideally, though, install a hose in the greenhouse for year round use, with an inside faucet to attach it to. Incorporating such an arrangement into the design of your greenhouse will save a lot of time and work. Also, I recommend buying an attachment that will allow you to spray a fine mist, beneficial for newly planted seeds and young seedlings.

Slow watering is always preferable to quick, forceful watering. The soil is better able to absorb the moisture and you'll get a true idea of how much the ground needs and can use. A quick flood will not penetrate as deeply to the roots and more is lost through evaporation.

Air circulation. Plants need good air circulation. In the summertime this is aided by having the exterior door and all exterior vents open. On winter days open vents to the house. Always keep fresh air circulating through the greenhouse.

Crop Rotation. Rotating crops is a highly recommended method of maintaining healthy soil and getting a good harvest over a number of years. Different crops take varying amounts of nutrients out of the soil; some actually enrich it.

The "heavy feeders" include tomatoes, cabbage (and members of that family), corn, all leaf vegetables such as chard, lettuce, endive, spinach and celery, cucumbers and squash. They need to be well fertilized and should *not* be planted year after year in the same bed.

Legumes (notably peas and beans, including soybeans and lima beans) are good soil builders; they "fix" nitrogen from the air, storing it in the soil for plant use. By alternating crops of legumes with the heavy feeders, you will be helping the soil to maintain a healthy balance of vital elements.

A third category consists of crops called "light feeders." Herbs and root vegetables belong in this group. If you're uncertain about how to classify a plant, it's usually safe to include it here.

Flowers break down into these same divisions. But unless you put all of the beds into flowers, they should not present a big problem. Just relocate them from time to time and keep the soil well fertilized.

Companion planting. Vegetables and flowers grown side by side compliment each other and make the greenhouse more attractive. Companion planting, mixing one plant with another in the vegetable row or flower bed, is good practice in any garden but is especially beneficial in the greenhouse.

Some reasons for companion planting are: certain plants excrete a substance above and/or below the ground which protects the plant—other plants growing nearby also benefit; a plant that needs a lot of light may be a good companion to one that needs partial shade; a deep rooted plant may benefit a shallow rooted plant by bringing up nutrients from deeper soil layers. Companion planting increases diversity in the greenhouse and tends to frustrate insect feeding and thwart buildups of insect populations. The Yanda's plant lettuce, radishes and other low growing vegetables under their tomatoes, peppers and eggplants. They plant clumps of herbs such as basil, anise and coriander throughout the greenhouse. Garlic and marigolds are found in each bed. Frances Tyson keeps garlic growing not only in many spots in her greenhouse, but also ground up in a jug of water in her refrigerator so that she can spray any white fly, cabbage moth, aphid, etc., which dares enter her greenhouse.

Be careful when planning the layout of your crops; companion planting does take some thinking through. There are obvious mixes that will not work—two plants that grow tall, two that have deep roots and so on. Some plants do not like other plants. For example, kohlrabi and fennel should not be planted next to tomatoes. Some plants that do well together are carrots and peas, kohlrabi and beets, tomatoes and parsley, celery and bush beans, radishes and lettuce, turnips and peas, onions and beans. Aromatic herbs make good companions as border plants.

DOGS AND CATS

Pets can be a nuisance in the greenhouse. Dogs with their bounding energy can trample seeds and plants into oblivion. They also love to dig into and lie down in a moist bed on a hot summer day. Cats consider the greenhouse beds a custom made kitty litter box. In general, all the problems you are likely to have with dogs and cats in an unfenced garden exist in the greenhouse. A gate or screen door is usually sufficient to prevent animal damage.

INSECTS

Joan Loitz, owner of The Herb Shop and Solar Greenhouse in Santa Fe, New Mexico, (see Chapter VII) has prepared a *Bugs in Your Greenhouse* manual which we are reproducing here:

BUGS IN YOUR GREENHOUSE

Greenhouses are special places—good for you, great for plants and of course potentially suitable for undesirable pests as well. Have no doubts that in time they will find you! You could build a greenhouse in the middle of a desert and...one day the slugs will appear. You could have a most tidy greenhouse, but a summer breeze could carry aphids or white fly in to take up residence on your tomatoes.

WHAT TO DO?—TWO STEPS

1. ***Prevention* is the best cure:**

Maintain proper growth conditions for plants. When *we* don't get proper rest, food and exercise, we are more susceptible to infection, flu, disease, etc. The same is true for every living organism, your plants as well. Be sure to provide them with proper water, fertilizer and temperature conditions and they will be able to stand off bug pests more easily.

Steady observation. Make a habit of checking plants regularly. Keep on top of any pest problems; they are much easier to control when not widespread.

Maintain proper ventilation. Interior air movement produces stronger plants and deters many insects.

Sanitation. Periodically remove dead, decaying or infested matter from plants. Keep floor and bench areas clean. When reusing potting containers, be sure to wash thoroughly and sterilize for one hour in a mild bleach solution (10 parts water: 1 part Clorox).

2. *Take Action* **Immediately. When you spot some undesirables, what are your choices?**

A. **First, there is topical treatment, i.e., sprays, dusts, etc.**

(1) **"Organic"**: Some of the suggested "organic" sprays, such as garlic and pepper juice, have very limited success in greenhouse conditions.

(2) **Botanicals**: These are naturally occurring pesticides. They biodegrade rapidly and have little residual effect. Nicotine, rotenone and pyrethrin are the major botanicals. Very suitable for use on edible plants. NOTE: Nicotine is fine as an insecticide but *don't* let smokers touch your tomatoes or smoke in the greenhouse. Tobacco mosaic is a tomato disease transmitted by touch. Smokers—wash your hands before fondling tomatoes!

(3) **Chemicals**. First there is the heavy duty stuff. Usually will produce quick results but do not degrade rapidly and have a high residual value. Chemicals are not recommended, especially for greenhouses that are attached to residences.

Then there are growth regulating agents. These are sprays which interrupt the life cycle of a plant pest, thereby inhibiting their ability to reach maturity and reproduce. They are just beginning to appear on the market.

Systemic insecticides are chemicals that are mixed with water and then fed to the plant. The poison then permeates the entire plant. When pests take a bite—they croak. Obviously, this is only for use on ornamental plants.

Techniques for spraying include hand sprayers (like an old window cleaner bottle) which are great for small jobs. One to two gallon pressure pump sprayers are better for larger jobs. When in doubt—DUNK!

Treatment mixture

Put a plastic bag around the pot—fasten at stem.

Turn plant upside down into treatment mixture, agitate gently, remove

CAUTION! Remember, all pesticides, botanicals as well as chemicals, are ***poisons***. Use only the recommended dosage or less. Keep children and pets out of the greenhouse for a safe period after their use. Use a mask if possible and gloves during application.

B. Another choice is **biological control**: using bugs that like other bugs for lunch.

(1) **General Note**: Biological control attempts to keep both beneficial as well as harmful insects in balance. Pests are only considered as such when their numbers are out of control. Predators actually feed on their prey. Parasites seek to complete their life cycles on or inside a specific host.

They are very effective in greenhouses! The use of predators and parasites for pest control was very widespread in the 1920's and 1930's in England and Canada. When World War II came and with it DDT and all the other petro-chemicals, we unfortunately abandoned many of the older and saner practices. We got seduced into believing these new methods were always better. As we all know now, the price to our environment and quality of life has been very high. Many plant pests have also been able to develop immunities to sprays. Thus, stronger chemicals have been developed; the cycle is indeed dangerous.

Predators and parasites are simply released to munch away an infestation. There are various types.

(2) ***Broad spectrum* predators and parasites**: lady bugs, praying mantis, green lacewings, trichogramma wasps, etc.

These are nature's own and always good to have around. *But* they have limited effectiveness in greenhouse situations. They're called broad spectrum as most of them prefer a smorgasbord of insects for lunch. They like a little of this and a little of that. Also, they are highly mobile. You can go through all kinds of efforts to bring in hundreds of ladybugs to your greenhouse, then after three days you won't be able to find any more around—because they left to find something else for dessert.

(3) **Frogs, Toads, Salamanders**

These are great to have around. They will eat big spiders and the grasshoppers and crickets that will just love to call your greenhouse home. They must be removed if you use any pesticides.

(4) ***Host specific* predators and parasites. Very effective in greenhouses!**

Nature's own—these are insects, usually quite small, that prefer only one or two types of bug for lunch. They are not as mobile and as such, are more effective in greenhouse situations. There are predators *or* parasites for: spider mite, white fly, scale, flies, mealy bugs.

(5) **Maintenance of predators**: Once a colony is released, the problem arises that they will die of starvation when they have eaten all their host. So to sustain a colony keep a "sacrifice" plant on which you allow a particular infestation of pest to continue. Periodically remove a few of the leaves and place them as food in areas where you know the predators are still at work.

C. **Here are some mechanical methods of pest control.**

(1) White fly, for example, love the color yellow. So, by hanging common yellow fly paper next to tomato plants, then shaking the plant, you can catch quite a few.

(2) **Cold fumigation**: most plant pests cannot survive a week without food. Thus, if you pick an appropriate time when there might be a lag in your planting schedule during the winter, just open the doors and let the greenhouse freeze out for a week.

(3) **Thumb and forefinger squash** (if you're squeamish, wear gloves)

(4) **Vacuum cleaner inhale** (good on white fly)

PEST	DESCRIPTION OF PEST	DESCRIPTION OF PLANT DAMAGE	TOPICAL TREATMENTS	NATURAL PREDATOR OR PARASITE	REMARKS
WHITE FLY	1/8'' LONG, SMALL WHITE: WINGED. QUITE OFTEN PRESENT IN LARGE NUMBERS ON THE UNDERSIDE OF LEAVES, WHEN THE PLANT IS DISTURBED THEY FLY OFF IN A CLOUD.	WHITE FLY SUCK PLANT JUICES AND CAUSE LEAVES TO YELLOW AND DROP OFF	PYRETHRIN ROTENONE REPEAT WEEKLY UNTIL CLEAR	ENCARSIA FORMOSA (PARASITE) EFFECTIVE IN GREENHOUSES	WHITE FLY WILL REPRODUCE MORE RAPIDLY WHEN NIGHT-TIME TEMPERATURES ARE LOW. WATCH OUT FOR ANTS.
APHIDS	1/16'' LONG. SMALL, SOFT BODIED, PEAR SHAPED SUCKING INSECTS. CAN BE BLACK, GREEN, YELLOW, BROWN OR GREYISH IN COLOR	THEY CONGREGATE ON THE YOUNG SUCCULENT GROWING TIPS AND FLOWER BUDS CAUSING STUNTING, CURLING AND PUCKERING OF LEAVES	NICOTINE PYRETHRIN ROTENONE	GREEN LACEWINGS (CHRYSOPA CARNEA)	FOR A SMALL INFESTATION, REMOVE AND WASH WITH MILD SOAP SOLUTION. WATCH OUT FOR ANTS.
MEALY BUG	1/4'' - 1/2'' LONG, COVERED WITH WHITE FUZZ WHICH MAKES THEM LOOK LIKE PIECES OF COTTON. THEY HAVE TAILS AND MANY LEGS.	WILL CAUSE NEW LEAVES AND FLOWERS TO DEVELOP POORLY. FOUND ON UNDERSIDES OF LEAF, AT LEAF AXILS & ALONG VEINS.	PYRETHRIN ROTENONE	CRYPTOLAEMUS	REPEAT TOPICAL TREATMENTS EVERY WEEK UNTIL CLEAR. ISOLATE PLANT. WATCH OUT FOR ANTS
SCALE	1/8'', WINGLESS, LEGLESS, WITH A WAXY, SCALE-LIKE COVERING. THE LARGER ONES ARE GREY OR BROWN MOTTLED WITH BLACK.	THEY MOVE VERY LITTLE SO OFTEN GO UNDETECTED. FOUND MOSTLY ON FERNS, CACTI, CITRUS AND SUCCULENTS ON STEMS AND UNDERSIDE OF LEAVES. WILL STUNT PLANT GROWTH	PYRETHRIN ROTENONE NICOTINE	APHYTIS MELINUS	ISOLATE PLANT. FOR MINOR INFESTATIONS WASH OFF WITH TOOTHBRUSH WITH MILD SOAP SOLUTION.
ANTS	YES, ANTS ARE A PROBLEM IN GREENHOUSES, PRIMARILY BECAUSE THEY ''HERD'' APHIDS, MEALY BUGS AND SCALE BY MILKING THEM FOR THE SWEET HONEYDEW THEY SECRETE. IF ANY OF THE ABOVE ARE PRESENT, CHECK ALSO FOR ANTS (ESPECIALLY IN SUMMER AND FALL). IF YOU CAN CONTROL THE ANTS, CHANCES ARE THE OTHERS WON'T BE AS MUCH OF A PROBLEM.				
SPIDER MITE	1/50''. VERY SMALL. BODY IS OVAL, YELLOWISH COLOR WITH 2 DARK SPOTS ON BACK (ADULTS)	MITES MAKE COBWEBS ON UNDERSIDES OF LEAVES AND FROM ONE LEAF TO ANOTHER. OFTEN COVERING A NEW SHOOT. CAUSE YELLOW PIN SPOTS ON LEAF	PYRETHRIN ROTENONE	AMBLYSEIUS CALIFORNICUS VERY EFFECTIVE	ISOLATE. CAN BE VERY SERIOUS PEST. ALSO THEY HAVE A HIGH RESISTANCE TO CHEMICAL SPRAYS, MOST ACTIVE IN WARM TEMPERATURES.
SLUGS	1/4'' - 1'' AND CAN BE MUCH LARGER. SMALL SLUGS ARE DARK GREY OR BLACK. LARGER ONES ARE GREY OR BROWN MOTTLED WITH BLACK.	THEY HIDE IN DARK, DAMP PLACES DURING THE DAY AND COME OUT TO FEED AT NIGHT CHEWING LARGE HOLES IN LEAVES.	SLUG BAITS, POWDER OR PELLETS. THEY HATE TO CRAWL OVER ANYTHING SHARP SO A CIRCLE OF CINDERS, LIME OR SAND AROUND A PLANT HELPS		SLUGS CAN BE A SERIOUS PROBLEM IF LEFT UNCHECKED. BE CAREFUL WITH SLUG BAITS--THEY CAN BE HARMFUL TO CHILDREN OR PETS. IT IS BEST TO PUT BAITS UNDER PIECES OF BOARDS, TIN CANS OR LIDS. SMALL CONTAINERS OF BEER OR GRAPE JUICE WILL ALSO ATTRACT SLUGS--EFFECTIVE THOUGH FOR TWO TO THREE DAYS ONLY.
GNATS	1/10''. VERY SMALL, LOOK LIKE SMALL BLACK FLIES.	SCAVENGERS. FEED ON DEAD AND DECAYING MATTER AT SOIL LEVEL. FLY UP WHEN DISTURBED.	PYRETHRIN ROTENONE		NOT A SERIOUS PROBLEM ALTHOUGH LARVAE WILL FEED ON SMALL PLANT ROOTS. IF YOU USE FISH EMULSION FERTILIZER THEY WILL BE MORE PREVALENT.
SOWBUGS OR PILLBUGS	1/4''. HARD OVAL SHAPED BODIES. USUALLY GREYISH. WHEN DISTURBED, THEY ROLL UP INTO A BALL, HENCE THE NAME ''ROLY-POLY.''	SCAVENGERS. FEED ON DEAD AND DECAYING MATTER. CAN HARM YOUNG ROOTS OR TENDER SEEDLINGS.	SLUG BAIT		SAME AS ABOVE. GOOD SANITATION AND VENTILATION HELP.

D. **Companion Planting:** Using plants that have a "bad taste" to repel pests from plants they find appealing. Marigolds, onions, garlic, chives, etc., planted near tomatoes tend to repel aphids. Many herbs such as sage, rosemary, thyme, rue, tansy and wormwood are also effective.

E. **Trap Plants:** Growing a specific plant to "trap or entice" a specific pest. For example, white fly is drawn to tomatoes like bears to honey. Plant one in your greenhouse–when it's loaded with white fly–gingerly remove it–dump it in a sack–then destroy.

F. **Genetic Resistance:** Many plants have been developed with built in resistance to various plant diseases and pests. Be on the lookout for these varieties in your seed catalogs.

Want to have some fun? Get a 10 power lens. Start poking around looking at leaves and bugs–up close–they're fascinating and kids love it!

Reference Books

Most plant books will have sections on bugs, but 2 *very good* knowledgeable references are:

The Gardener's Bug Book (4th edition)
by Cynthia Westcott, Doubleday & Co., Inc., N.Y.

The Encyclopedia of Organic Gardening
Rodale Staff, Rodale Press, Emmaeus, Pa.

Sources of Supply:

Topical Insecticides: Most local nurseries and plant shops.

Biological Control:

Biotactics (predators for spider mite)
22412 Pico St.
Colton, CA 92324

Rincon-Vitova Insectaries (predators for spider mite, scale flies, mealy bug) Also trichogramma, lacewings.
Box 95
Oak View, CA 93022

The Herb Shop (predators for spider mite, scale, white fly)
1942½ Cerrillos Road
Santa Fe, N.M. 87501

Bio-Control Co. (ladybugs)
10180 Ladybird Ave.
Auburn, CA 95603

PEOPLE AND PLANTS

When children are old enough to contribute their love and labor to the greenhouse, I urge you to let them. The only skill required for planting, watering and harvesting is judgement—and perhaps intuition.

FIGURE 49

In my experience, children love to work with plants and the plants seem to benefit from the care of children. The results are fairly immediate and highly visible, yet mysterious. How does a seed turn into a plant? But it does and it grows quickly. Children have a single-minded kind of energy that can be channeled into extraordinary care for the plants. It's a mutually supportive relationship.

The satisfaction of greenhouse gardening extends to people of all ages. Most of the work is not strenuous and its rewards are great enough to encourage everyone to give it a try. The natural pace of this work is unhurried; the environment is relaxing and absorbing. You can enter the world at a different level in the greenhouse, be transported and totally involved at the same time.

CHAPTER VII The State of the Art

In this chapter we'll give you an idea of what's being done in the field of solar greenhouse design. As you will see, there is no *one* solution to combining greenhouse features. These structures, which are only a representative selection of the kind of work that's going on, offer an amazing variety of approaches. Most of the designers and owners are happy to share their knowledge and experience; we urge you to contact them for more information. When writing them, *always* enclose a self-addressed stamped envelope. Many sell detailed plans and offer consultant services. You will notice that the designs include freestanding units, attached or retrofitted models, and newly built greenhouse/home combinations.

We have divided this section up into individuals, research institutions and organizations, and manufacturers. Sometimes they overlap.

INDIVIDUALS

The Herb Shop

This unit might be considered the mother of solar greenhouse design (see Figure 50). As far as I know, it was the first such commercial greenhouse in the United States. The design is based on the work of the Brace Institute (in Canada) and T.A. Lawand. Joan Loitz, the owner, adjusted the angle of the front face and the reflective back wall to correspond to the latitude of Santa Fe, New Mexico. The north wall is at a 78 degree angle (angle of the sun at summer solstice) and is insulated with 4" of fiberglas. The interior is paneled with Masonite and painted with a glossy white enamel. The rafters extend from the roof peak to the ground and were originally uncovered for the lower 10 feet. Joan covered this area with corrugated Lascolite to make a greenhouse "preheater" or "buffer zone" and cold frame area. It stays about 25 degrees colder than the main greenhouse at night. This addition cost her 90 cents per square foot and gave her an extra 480 square feet of growing space.

Occasionally a customer will challenge Joan about her use of supplemental gas heating. She replies that she's in the business, not playing games, and that the heater is the most cost-efficient way of backing up a system that's 80 percent solar. Joan estimates that it would cost her $3,000 to install an active solar system to capture the 20 percent nonsolar power that the gas company furnishes. Her average $32 a month gas bill for the eight-month cold season compares to $160 a month for a similarly sized conventional greenhouse in the area, and bears out the effectiveness of her design. Incidentally, she's running 70 degree nighttime lows, which is considerably higher than those maintained in noncommercial greenhouses.

FIGURE 50

Some thermal storage is provided by sixteen 55-gallon drums that also double as table supports. Although these drums receive little direct sunlight, they do contribute to heating when the ambient air drops below their temperature and help to stabilize temperature fluctuations.

One drawback to the design of this unit is that the entire clear area approximates a normal angle to the summer sun. This may be all right for Quebec, but it creates overheating and plant-burning problems in Santa Fe. Joan has used some ingenious low-cost methods to combat this problem. One of them is to attach a $1.98 misting nozzle to a hose and hang it in front of the air circulating fan. This will lower the temperature 10 degrees in 10 minutes; she turns it on several times during a summer afternoon. She also trains scarlet runner beans up the west wall to block some of the afternoon sun.

Joan and I have often worked together in planning community solar greenhouse systems. Her particular expertise is in organic growing methods and biological control of insect pests (see Chapter VI). She ships 100 varieties of organic solar-powered herbs all over the world. You can order from the address given below.

Owner: Joan Loitz, The Herb Shop, 1942½ Cerrillos Rd., Santa Fe, NM 87501

Designers: Brace Institute of McGill University, and Ms. Loitz

Builders: Joan Loitz and friends

Floor Area: 1600 square feet

Clear Area: 1042 square feet (greenhouse); 480 square feet (cold frame) = 1592 total square feet

Thermal Storage: Sixteen 55-gallon drums (880 gallons of water)
Supplementary Heating: 150,000 BTU (adjusted) natural gas heater
Material Costs: $3.40 per square foot, plus beer (1974)
Main Function: Wholesale, retail and mail order sale of organically grown herbs

Wesley and Frances Tyson

I designed and built the Tyson's solar greenhouse for them shortly after they moved from New Jersey to Santa Fe. After examining their property, I told the Tysons that there was no way to attach the unit to their home without demolishing several trees. Frances didn't really care, as she considers most of the large imported trees surrounding her home water mongers anyway. But I didn't want to be the assassin. So, we decided to build an independent greenhouse on a hillside near the home. At that location the unit could be sunken about 4 feet into the hillside and the north wall earth bermed (earth piled against the wall above grade).

This unit has proven to be the best in a long series of my designs and contains several important innovations that seem to work well together. The tilted back wall (after the

FIGURE 51

Brace Institute design) guarantees that the plants get sufficient north light (reflected) all year. The wall is a tight sandwich of insulating materials. From the inside out it contains: ¼" sheetrock, textured and painted glossy white, aluminum foil, 6" fiberglas insulation, ¾" Celotex siding, 6-millimeter black polyethylene, and 1" overlapping rough lumber (Figure 51).

The south wall is at a 75-degree angle to the horizon and is framed up on a low (16") concrete-filled, pumice block wall. The southern skin is approximately 110 square feet. The south-sloping roof has alternating clear (fiber acrylic) and opaque (tin, insulation, sheetrock) panels. A roof vent is set into one of the solid panels and is opened by an automatic heat piston.

The roof may be the key to the successful performance of the structure. My brother-in-law, Paul Bunker, collaborated with me on the design. Because alternate sections of the roof are insulated, the unit stays warmer in the winter and much cooler in the summer. The solid roof sections are also perfect for mounting hinged moveable panels if extra insulation were needed. In this greenhouse there is only 50 square feet of clear roof approximating normal to the summer sun.

Another important factor in the design of the Tyson greenhouse is that it is oriented 25 degrees southeast of due south. This angle was determined mostly by the lay of the land and our desire to alter the natural environment as little as possible. Since very few morning rays would enter the greenhouse through the east wall at this orientation, the east wall is 70 percent opaque and insulated. The inside of this solid wall is reflective (tomatoes that are situated directly in front of it show excellent blossoming and fruit set). The west wall is clear.

This greenhouse was designed to contain ten water drums and have moveable insulating panels on the roof. With these provisions, I estimated that the unit would maintain 50 degree interior temperatures at -10 degrees outdoors. With only four drums installed and without the styrofoam panels in place, the greenhouse held a 43 degree low at -4° outside. That's truly remarkable and I can only attribute it to the roof design and the mass of the insulated block wall (it has an inch of styrofoam on the outside surface).

Frances Tyson is an experienced gardener. For years she had a large organic garden back in New Jersey. Frances uses growing techniques that would make a chemically oriented gardener turn over in his sodium nitrate. She uses relatively "hot" manure, heavy green mulch, never sterilizes anything and plants extremely densely. I've never seen a greenhouse that could match the level of production she gets out of the space she has. Some of Frances' techniques are discussed in Chapter VI. This amazing woman also finds time to build solar collectors out of beer cans, and to speak to anyone who will listen about the development of solar energy and the abolition of nuclear power. (See Figure 52.)

Owners: Wesley and Frances Tyson

Designer: Bill Yanda

Builders: Paul Bunker, Susan and Bill Yanda

Floor Area: 180 square feet

Clear Area: South face = 110 sq. ft.; south roof = 50 sq. ft.; east wall = 25 sq. ft.; west wall = 50 sq. ft.

Thermal storage: Pumice block walls filled with concrete = approximately 7000 pounds; 220 gallons of water (enclosed drums)

Material Costs: $5.00 per square foot

Main Function: Vegetable production for personal use.

FIGURE 52

William and Katherine Otwell; Michael Frerking

Bill and Katherine Otwell live in a passive greenhouse/house combination that Bill and Michael Frerking designed (see Figure 53). The amazing thing about their home is that its total material cost was $1,500, and that includes a solar hot water heater. It took the Otwells and Frerking three months to complete the 650 square foot structure. The walls of the home are raw adobe bricks, made on the site. (In more humid parts of the country stabilized adobe should be used.) The walls are insulated with 1" of styrofoam on the out-

FIGURE 53

side and plastered over with stucco. The greenhouse collector is made from recycled glass, and the roof is recycled billboards. At night they cover the glass with 2" of styrofoam for insulation. There seems to be difference of opinion concerning the amount of time required to put the styrofoam cover on the glass, three minutes to 15 minutes. Either way, I think 10 degrees in temperature savings is well worth the installation time.

The Otwells use a combination of sewer sludge, sand, blood meal and potting soil for a planting medium. They have misted plant leaves with water to control their main insect problem, white flies. Katherine has noted that a greenhouse entirely integrated with the home poses one problem in particular—insect hassles are brought into the house and shared with the plants there. In that respect, an attached solar greenhouse that can be shut off from your home has a certain advantage.

The Otwells' greenhouse/home combination had an average low of 55 degrees and average high of 75 degrees during the first winter in use. Corresponding outdoor temperatures were 10 degrees and 50 degrees respectively.

So there they sit, Katherine and Bill Otwell, in a beautiful low-cost completely passive solar home in the Arizona desert. They're eating fresh vegetables from their greenhouse when it's below freezing outdoors, burning a minimum of wood in very cold weather, designing other low-cost solar homes, and defying the notion that such dwellings don't exist yet! For further information write to: William Otwell, Arizona Sunworks, Star Route, Chino Valley, Arizona 86323.

Jemez House Boys' Ranch

The Boys' Ranch, located near Española, New Mexico, is a place where boys with learning disabilities or unstable family conditions are cared for and educated. The man who has had a great deal to do with the development of the large solar greenhouse on the property is Bob Detwiler. Bob has had six years of experience in greenhouse operation and

FIGURE 54

management. He is also a professional counselor for the retarded and for young people with learning problems. He sees the greenhouse as a natural rehabilitation device: "Everyone can profit from working with plants. I think one of the reasons for this is that there is a kind of natural feedback mechanism involved. When the plants do well, they show us the right way of working with them. In this way the plants validate our innate intelligence."

The original design for this freestanding greenhouse was conceived by Jay Davis. It employed a system of exterior insulating panels and shading devices that Bob felt was too complex, so he asked the Sun Dwellings team to modify the design. They incorporated most of the features of Davis' plan, but they eliminated the panels. The Sun Dwellings designers decided they could create an acceptable year-round growing environment by increasing thermal storage in the rock beds and adding a simple system of fans to tap the apex heat and blow it down through the rock storage. Fidel Lopez, the manager of the physical plant at the Ranch, further amended the angles of the roof to produce a basic 45 degree A-frame structure. This configuration will begin to shade the north wall at solar noon on the spring equinox.

FIGURE 55

The active heat transfer system uses four 10" ¼-horsepower fans. The ducting is 10" reduced to 8" galvanized steel stovepipe (see Figure 55). Excess heat is exhausted through three 18" roof vents. More venting is needed, however, and Bob plans to add low west and high east vents in the walls.

For the first winter's operation, the active air circulation system was not completed. Even without it, the boys began planting at the end of January and produced a healthy crop of petunias, cabbage, tomatoes, and chili. With only passive direct-gain storage in the rock beds, 45 degrees was the lowest temperature recorded, with a fifteen degree outdoor low. Bob has estimated that with the active system in operation, on a good solar day in winter approximately one million BTU's will be vented from the unit to help heat the frosty air of the upper Rio Grande valley. That's one disadvantage of an independent greenhouse.

The Boys' Ranch looks forward to next winter, when the greenhouse will put a full array of fresh vegetables on the dinner table.

Owner: Jemez House

Designers: Jay Davis, Bob Detwiler, Sun Dwellings Design team, and the Yanda greenhouse class.

Builders: Bob Detwiler, Fidel Lopez, boys and staff of Jemez House, Los Alamos Civitan Club.

Floor Area: 900 square feet (floor 5 feet below grade).

Clear Area: 720 square feet (front face).

Thermal Storage: 70 tons of rocks (assorted sizes) and earth berm.

Insulation: Perimeter = 6" of sawdust in plastic envelope; roof = 6" fiberglass; door = 2" styrofoam and 1" fiberglass

Material Costs: $3.88 per square foot, $3500 total

Main Function: Supplementary food production, retail sale of starts and bedding plants.

Hamilton Migel

Hal Migel is a designer and builder of exceptional talent. I first became aware of his professional skills seven or eight years ago when every adobe mudslinger I knew was saying,

FIGURE 56

"have you *seen* the detail work in Migel's newest house?" Attention to detail, function and form personifies his work.

Migel was the first person in this region to demonstrate that a greenhouse could be the primary heating unit for a large home (2500 square feet). He did it with some revolutionary innovations.

The double-walled clear face is separated (or inflated) by a small fan and is supported by thin steel ribbing. The entire clear wall can be detached from the greenhouse front in late spring, leaving a lovely outdoor garden in full bloom. Removing the cover also provides a clear view to the outside (also see Solar Room, p. 137).

Warm air is ducted from the greenhouse apex to a rock storage bed under the house, then back into the unit to complete the loop. Radiant heat from the concrete slab over the rock bed warms the house. The system provides approximately 80 percent of home heating.

FIGURE 57

Owner: Hamilton and Candice Migel
Designer: Hamilton Migel
Builders: Bingham and Migel, Inc.
Floor Area: 400 square feet (greenhouse)
Clear Area: 500 square feet (greenhouse)
Home Floor Area: 2500 square feet
Thermal Storage: Rock beds under floor of home
Air Circulation: 1100 to 2200 cubic feet per minute (see intake vents, upper left, Figure 57).
Main Function: Home heat, vegetables and flowers

David J. MacKinnon

Dave MacKinnon is not only a scientist with sterling credentials but also a visionary with the hard data to back up his statements. In conjunction with Vincent J. Schaefer of the State University of New York at Albany, Dave has written a very comprehensive paper entitled "Can Vegetable Productivity be Economically Increased in the Northeastern United States?" The following excerpt gives his economic rationale for home greenhouses and some spatial criteria for the designs.

> As part of a greenhouse research program conducted by the Atmospheric Sciences Research Center of the State University of New York at Albany, Rodale Press, Inc., and the Museum of Northern Arizona of Flagstaff, Arizona, a greenhouse is being developed for the purpose of providing the individual consumer with vegetables from the home garden. It is hoped that the greenhouse will not just supplement, but provide almost all vegetable needs. The greenhouse is planned to use little or no energy, other than solar, to maintain a plant growth environment year long in severe climates. A detailed cost analysis for the initial design of the greenhouse has shown that the materials will cost from $2.50 to $3.00 per square foot of growing area covered (retail prices in Flagstaff, Arizona). After material testing, it is likely the costs can be reduced $0.25 to $0.50 per square foot of growing area. Construction will be sufficiently simple to be performed by the home owner himself. The question is: How economical is it to grow your own with this greenhouse?
>
> Statistics for 1971 show that roughly 200 square feet of field space was required to provide the per capita yearly consumption of fresh and processed vegetables (considering only the most important vegetables: lettuce, onions, tomatoes, cabbage, celery, carrots, cucumbers and green peppers). The greenhouse can be at least 5 times more productive per acre than the field crop, thus requiring 40 square feet of greenhouse to feed the average person every year. Thus, the capital investment per person for the home greenhouse space is between $100 and $120. Taking the difference between the retail price for vegetables

purchased in New York and the average farm price of those vegetables (the farm price is probably higher than the actual home grown production cost) yields the markup the consumer must pay for vegetables. Multiplying the markup by the consumption rate for each vegetable gives the potential savings achieved by eating home grown, fresh vegetables; this amounts to $14.09 per person per year for all fresh vegetables in 1971. Since the markup for processed vegetables is at least twice the fresh markup, the use of home grown fresh in place of processed saves roughly an additional $28.18. A total savings of about $42 per person per year can be achieved. The greenhouse capital investment can be paid off in three years. The designed life of the greenhouse is at least 20 years, giving about 17 years of low cost, high quality vegetables. By going home grown the savings are not really spectacular, but then neither is the amount of vegetables the average person consumes. Perhaps when there is little else to eat, vegetables will be worth their weight in gold. There is no telling what the impact of developing a home grown technology will be—it may sink or swim, we'll see in the years to come.

FIGURE 58

FIGURE 59

The paper also presents major alternatives to a possible food disaster in the highly populated northeastern part of the United States.

Dave is also busy designing and building the very exciting solar greenhouse shown in the photographs above.

Douglas Davis

Doug Davis of Gunnison, Colorado, has designed a greenhouse appropriate to his area, using a water storage system heated by hot air collectors made of metal lath inside the unit (see Figure 60). The collectors are single glazed, with the greenhouse clear wall serving as

PHOTO BY C.F. KOONTZ FIGURE 60

a second insulating layer. Davis' design utilizes 2000 aluminum beer cans filled with water in a heat storage bin located under the growing area. A radiant floor keeps the interior warm at night. Moveable insulating/reflecting shutters also reduce nighttime heat loss and protect the glass from vandalism.

In the first winter of operation, the Davis greenhouse held temperatures in the high 40's in extreme outside conditions of -30 degrees.

Doug is presently putting together a book on energy conservation in building design, tentatively titled *The Heat-Tight House.* He also offers consulting services on active and passive solar heating systems. For more information write him at: 313½ North Pine Street, Gunnison, Colorado 81230.

James and Elizabeth DeKorne

Jim DeKorne is a familiar name in the literature of alternate energy research. He is a regular contributor to *Mother Earth News*, and Jim's latest book, *The Survival Greenhouse*, is an excellent source of information, particularly in the areas of hydroponics, wind generation and complex ecosystem design. Jim and Elizabeth have set for themselves the challenging task of building a self-sufficient life style on one acre of semi-arid, high altitude land in northern New Mexico.

FIGURE 61

Our main interest in Jim's work concerns his use of varied life forms and independent power in his greenhouse designs. He has found, for instance, that introducing rabbits into his pit greenhouse improves plant growth. He believes that the carbon dioxide that the rabbits exhale compensates to an extent for the short photoperiod in winter months. The DeKornes are also fond of rabbit for the table, and a pair of breeding stock will produce as much protein in one year as a large heifer provides.

Jim's system also incorporates a wind generator that, at one stage of experimentation, powered an aerator for edible fish that ate soil-aerating worms whose wastes were used to make nutrient solution for hydroponically grown vegetables...quite a system.

The only reservation that Jim has about his original pit greenhouse design is that it was not attached to his house. The excess heat vented in the winter could be put to good use. He plans to build an attached unit this summer.

I think that the varied approaches Jim has been involved with warrant further research. Hopefully some new greenhouse builders will be interested in developing complex ecosystems that will contribute to the existing knowledge in this area.

Edward and Barbara Storms

Ed Storms decided to build an attached solar greenhouse after attending the 1974 Ghost Ranch Bio-Technics conference at which I gave a talk on the subject. The greenhouse and a workshop were to be added to the frame house that he and Barbara had built. Since the addition would have no thermal storage capacity in its frame walls, Ed decided to include an active warm air system with rock storage under the unit.

Ed encountered many a do-it-yourself disaster as he worked on the project. He had a backhoe to do the excavating and it immediately hit the buried gas line. Then they came across a boulder that even the backhoe couldn't budge. They built around it. Ed hired a concrete company to bring over the mix. The next two hours were the biggest scramble of his life.

Completed now, the unit is a clean and attractive addition to the home (Figure 62).

FIGURE 62

The design work for the curved fiberglass/polyethylene face was done by Mike Watson, a Santa Fe architect. The outer fiberglass is overlaid with molding and bolted to the curved steel struts. The small greenhouse heats Ed's 12 x 25' shop addition. The Storms' installed a back-up wood stove that they haven't had to use yet. Heat to the shop is supplied by natural convection through the adjoining door.

The 16-cubic-yard rock storage pit has layers of sand, plastic, then Celotex separating it from the natural volcanic tuft at the bottom of the excavation. Hot air from the apex of the unit is moved down a black plywood duct and through the rock bed. The fan is activated automatically when the inside air temperature reaches 70 degrees. Ed's extensive monitoring of the temperature differentials in the rock storage bed indicates that the front (south) of the plenum is not getting enough warmth to substantially contribute to night heating. He believes that the temperature profile of the plenum would be higher and more even if he substituted a larger fan for the 6" squirrel cage blower now in use. Overheating in the unit due to the all-clear cover could pose a problem, though some high shading and cross venting should handle it quite easily.

The nighttime heating performance of this design (even without a larger fan *and* without insulation for the cover) is quite impressive. It has maintained a 43 degree low in -7 degree outdoor temperatures. This is good for Los Alamos, at an elevation of 8,000 feet. Ed does have 80 gallons of water in drums for passive, direct-gain storage (see Figure 63).

The Storms' are living a refreshing low-energy existence in the suburbs. They have some very creative designs for a small lake and trout stream operation that would utilize rain water captured from the roof of their home. They have also created a well-planned terraced garden to receive greenhouse starts, with a compost area to maximize organic food production in a very limited space.

FIGURE 63

Detailed plans of the Storms' greenhouse are available through: Energy Systems, 77 La Paloma, White Rock, New Mexico 87544.

Owners: Ed and Barbara Storms
Designers: Mike Watson, Ed Storms
Builder: Ed Storms
Floor Area: 140 square feet

Clear Area: Fiberglass exterior/602 interior
Thermal Storage: 16 cubic yards of rocks, 80 gallons of water
Insulation: 1" styrofoam around the perimeter and rock bin
Material Costs: $7.00 per square foot
Main Function: Winter vegetables; vegetable starts.

Lee Porter Butler

Lee Porter Butler, a San Francisco-based architect, has developed a design that combines natural air circulation features in a unique way. He calls the passive design a "gravity driven heating and cooling system." It is applicable to houses, larger structures and greenhouse/home combinations. The diagram in Figure 64 shows a schematic cross section that illustrates the total concept that Mr. Butler has designed into several homes (see Figure 65).

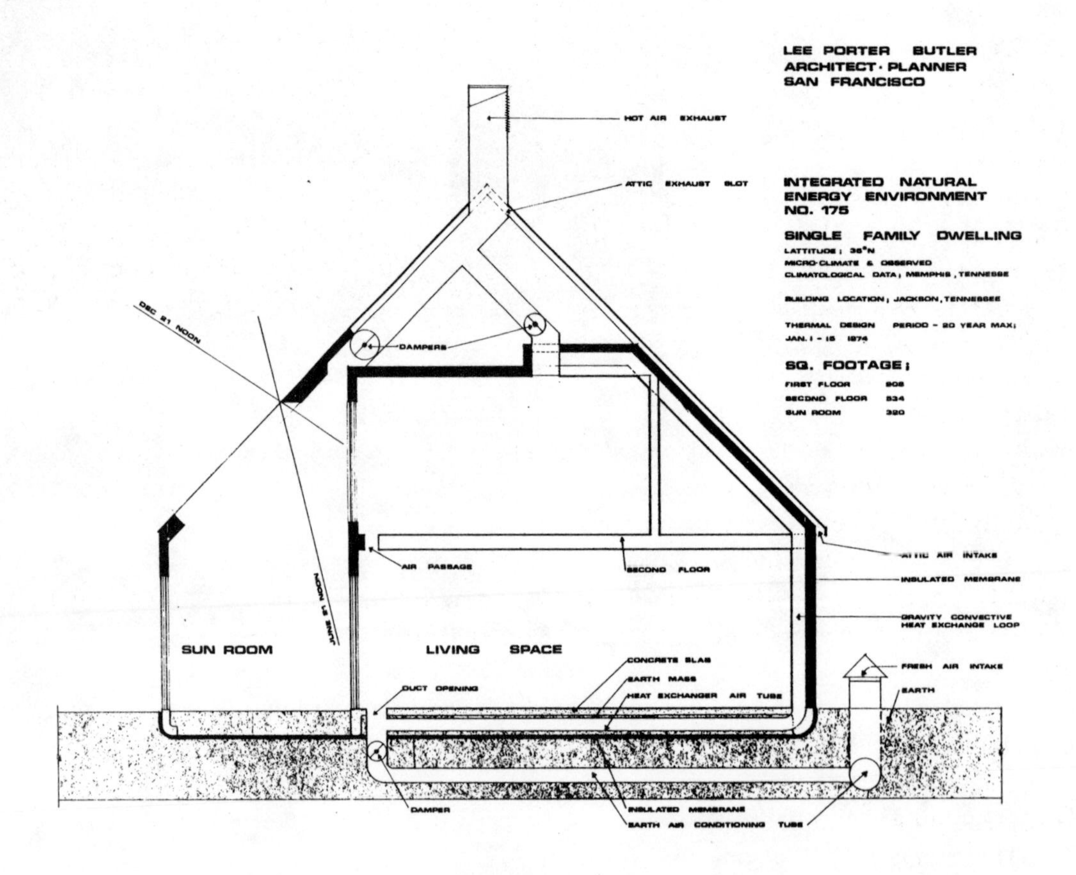

FIGURE 64

One interesting aspect of the plan is that the heat storage medium is common earth from the site. The most important feature of the design is the double roof and north wall construction. It is in effect a house within a house. Hot air from the greenhouse is pulled down and through the storage area by the cooler, heavier air that falls down the north side. This warm envelope surrounds the house for winter heating. In summer a roof vent is opened and outside air is pulled through the 55 degree earth to cool the greenhouse and home.

FIGURE 65

Lee first built a 4000 square foot home based on this design for his family while living in Tennessee. The first winter of occupancy a once-in-a-century ice storm hit, knocking out all power lines and driving the nighttime temperatures below zero. Even though the days were overcast, with temperatures never exceeding 34 degrees, the home held at 70 degrees during the day and didn't drop below 58 degrees at night (with no supplemental heat). Lee opened the top levels of the home to the large south-facing greenhouse and raised the temperatures even more.

Lee has dedicated the last several years to refining the design and substantiating his data. He emphasizes that the design demands a full understanding of the thermal principles that determine dimensions of ducting and construction of the storage facility and double wall.

Gregory Franta

Here's an ambitious plan for a totally integrated semi-passive dwelling in the alpine climate of Aspen, Colorado. Designed by Greg Franta of the Roaring Forks Research Center in Aspen, it utilizes earth berm and low profile techniques to conserve heat (Figure 66). The greenhouse has adjustable dark on top, light on bottom insulated louvres for collecting, shading and insulation (clear view type). The sod roof and natural overhang help make the structure a cool summer space.

FIGURE 66

Domestic hot water is supplied by a collector (60 square feet) above the bathroom. The water is circulated into a 66-gallon tank by natural thermosiphoning. A Clivius Multram composter is used for human and kitchen wastes. Back-up heating is provided by two wood-burning stoves.

The heated air from the greenhouse is blown down into rock storage beneath the first story and that floor receives radiant heat from the 50 cubic yards of rocks. The vents for air distribution are manually operated. (See Figure 67.)

In my opinion, this design combines many of the best features of semi-passively heated homes. The only major loss areas (gain areas) on the south side are all insulated manually. The bedrooms, where more night warmth is needed, are upstairs. The living, dining, and kitchen areas adjoin the greenhouse. It's quite a sensible design and one that depends on little outside energy to make it function.

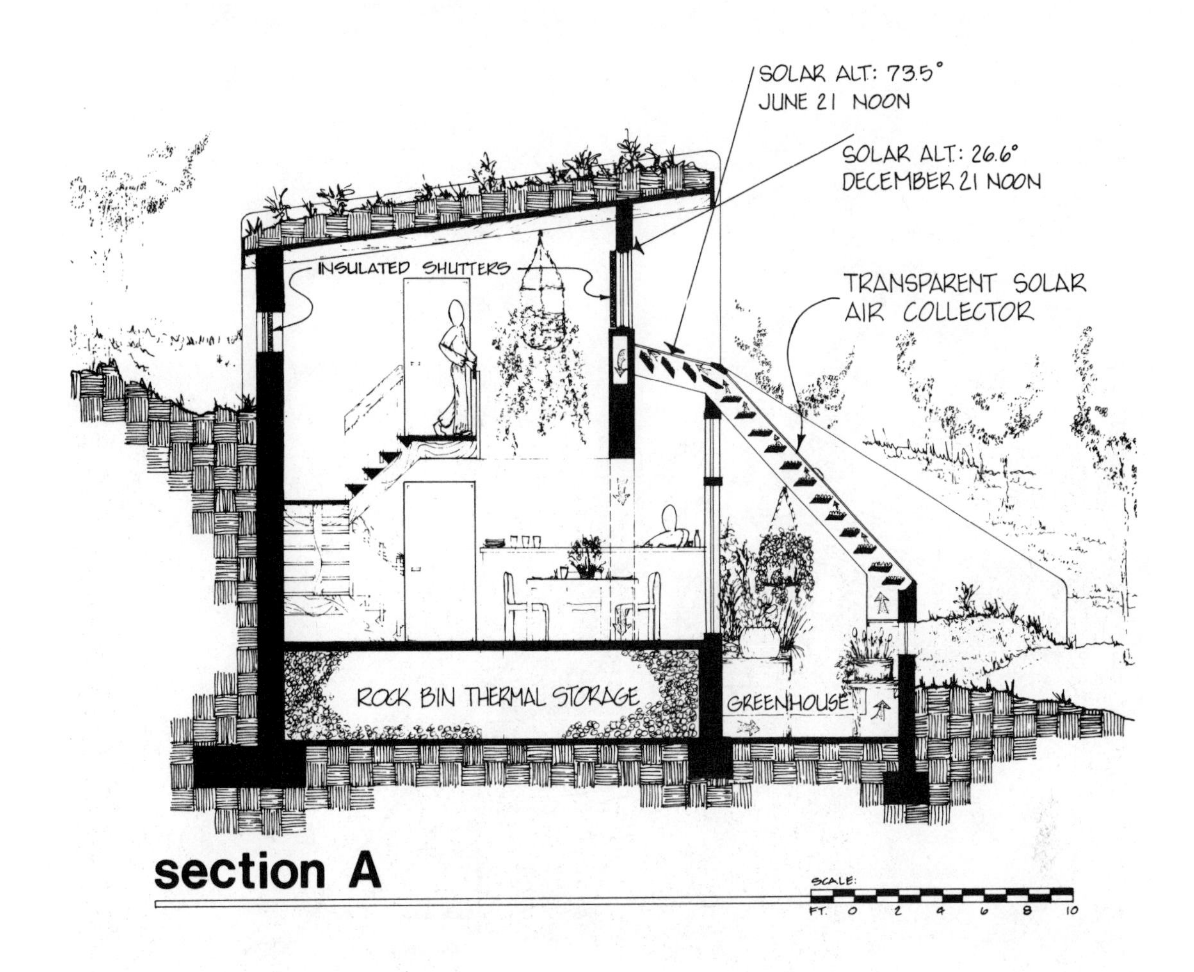

FIGURE 67

Lloyd Wartes and Tim Carpentar

An interesting design related to John Todd's earth covered biosphere (New Alchemy Institute, page 120) is being developed by Lloyd Wartes. Wartes is an agricultural engineer and designer with over fifty years of experience in a multitude of disciplines. Along with Tim Carpentar, a hydroponics expert from Colorado Springs, Colorado, he has designed a greenhouse that will maximize natural geologic contours in growing food. Lloyd and Tim expect to see several acres of near-vertical Rockies converted to food production in the coming years.

The terraced greenhouse will be built up the side of a steep grade (see Figure 68). Heated air will rise from one level to another, producing *different* environments on each level. Using for example three popular greenhouse crops, lettuce would be on the bottom, tomatoes in the middle and cucumbers in the warm top section. A Chinese kiln for plants: simple and brilliant. The same concept could be applied to your own backyard, providing you are perched on the edge of a precipice.

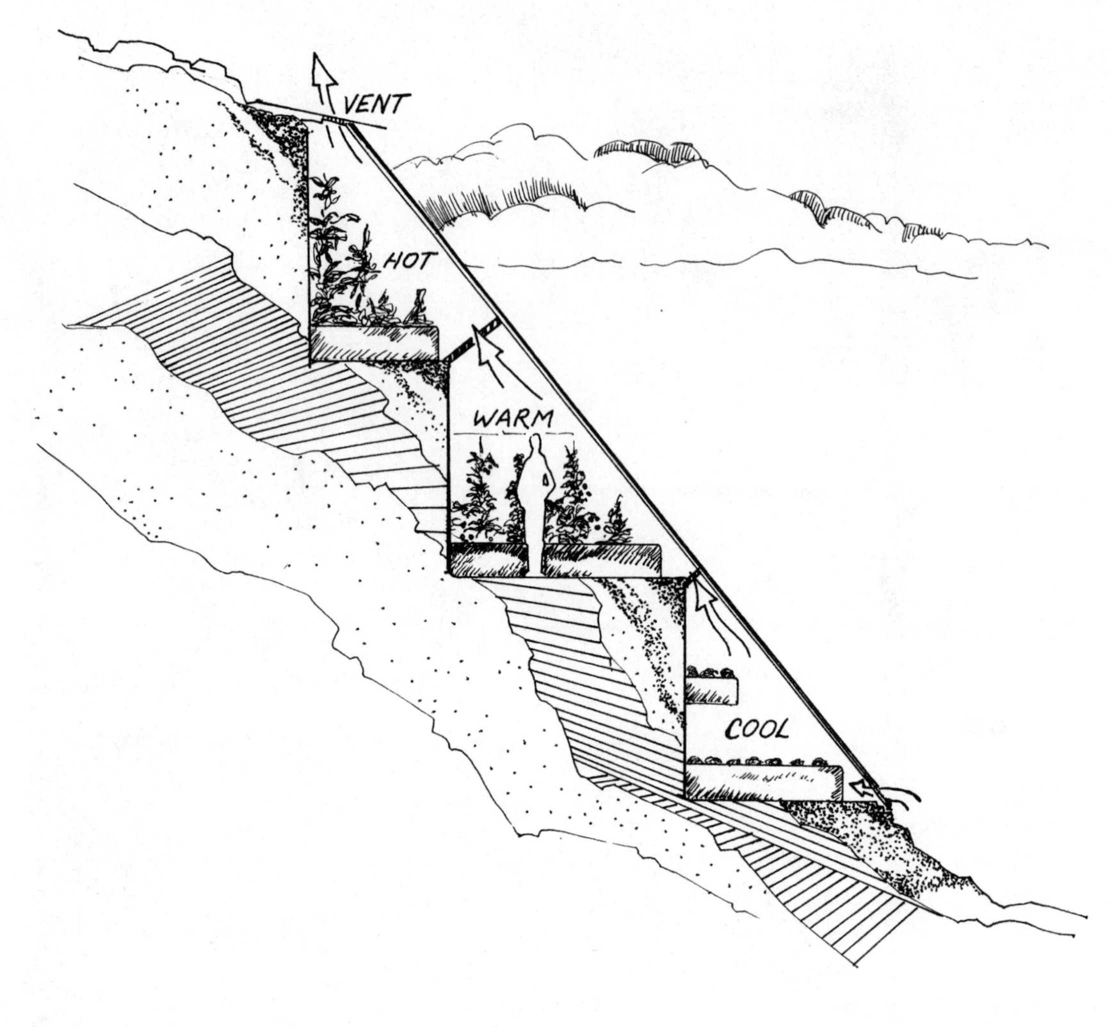

FIGURE 68

Barbara and Peter Voute

This unit has a configuration very similar to Dave MacKinnon's independent greenhouse seen on page 97, Figures 58 and 59. But as far as I know, they were designed completely independently of each other. The south face is at a 65 degree tilt and the clear roof section slopes away from the front (toward the north). The two clear panels running the length of the roof are used only in the summer for north lighting. In the winter they are covered with rigid styrofoam to reduce heat loss.

In designing this unit, I experimented with a glazed direct-gain wall. I think it's a good idea that needs further development. This is how it works.

FIGURE 69

The coldest area in the greenhouse is in the extreme south bed at ground level. I knew that if the earth there could be heated, the plants would do better through the long cold winter nights. The south face is framed up on a 16" high by 10" thick concrete wall. I painted the outside of this low wall dark brown and glazed it with one layer of fiberglass acrylic material. It was well sealed on the top and bottom. Heat is transferred through the concrete to the earth bed directly behind it. (This is like a Trombe wall but has no vents for air circulation.)

The results were that the young plants nearest the wall showed faster, healthier growth than those farther back in the bed. However, the stored heat wasn't enough to carry some of them through the coldest winter night. When the outside temperatures dipped to -7 degrees, a patch of beans in the front bed froze. Nothing else was damaged, including tomato plants situated centrally in the unit. But Dr. Voute was disappointed and he bought a small electric space heater calibrated to turn on at 45 degrees.

FIGURE 70

To improve the performance of the direct-gain wall I would: 1) make it thinner; 5" would be thick enough for strength and would provide a higher rate of heat transfer to the front bed; 2) double glaze the south low wall; this would prevent such rapid conduction losses back through the wall at night; and 3) insulate the lower clear wall (at night).

The Voute's grow some beautiful flowers and vegetables in this greenhouse. The soil mixture they use is 1/3 city sludge from the sewage plant, so occasionally they'll have a healthy tomato or chili plant pop up in an unexpected place. Dr. Voute has also developed a rather intense personal relationship to his unit. To paraphrase him: "The damn thing's like a spoiled pet or child; it needs attention all the time. I think we'll skip December and January growing next year and take a vacation from it." This is said lovingly, believe it or not.

Owners: Peter and Barbara Voute

Designer: Bill Yanda

Builders: Paul and James Bunker, Bill and Susan Yanda

Floor Area: 160 square feet

Clear Area: 128 sq. ft. (south face); 40 sq. ft. (east and west walls); 32 sq. ft. (roof)

Thermal Storage: 3 cubic yards of concrete in walls (approximately 12,000 pounds) and 330 gallons of water

Supplementary Heat: Electric space heater

Material Costs: $4.00 per square foot

Main Function: Vegetable and flower production for home use

Wayne Nichols, Seton Village, New Mexico

If you could combine the experience of a Harvard business degree, a year of concentrated meditation in Spain and Italy, several years working for a multi-national conglomerate in Los Angeles with the love of playing with mud houses and a natural bent for philosophy...what would you have? Wayne Nichols, of course.

Nichols is what I call a prime mover. He has the ability to centralize the talents of many individuals and direct their efforts into a tangible reality. He has earned the right to wax philosophical because his visions and dreams are based on hard-nosed experience. For instance, how many dreamers in the U.S. have conjured up solar villages and low energy communities? (usually funded with monopoly money). I've personally heard of about 5000. Well, Wayne has built one. It's out there in the piñons. You can buy a home in the community...today. Read these excerpts I've taken from a paper Nichols recently wrote. They make more sense than government feasibility studies and they didn't cost you a penny of tax money.

> "Man himself with his fossil fuel, technologically based society and high population levels is in for a crash. Could Americans really change if they wanted to? I doubt it. To many of us it would be impossible to adjust to a culture without our machines and their energy consumption. If we ran out of oil there is a good chance we would fight rather than alter our culture.
>
> In the end nature always wins. If we continue to act like spoiled children wasting our precious natural resources and despoiling our environment, Mother Nature will take corrective action. She always does.

FIGURE 71

FIGURE 72

We are a part of nature and we will carry out the orders for our own discipline.

...As technology has increased, so has our ability to destroy ourselves. To balance this, our ability to change has increased through media to control the expanding threat of technology. Technology is held in balance by media. It is such a simple formula: man in combination with his environment form a single system that is self-regulating, self-managing, automatic. As Steve Baer says: 'God is the original passive system.'

In America we can make changes very quickly. We have a relatively organic political structure, and a free market system that allocates resources (capital) very quickly, where we need them. Through the partnership of our incredibly powerful media and our scientific capacity America can change and change rapidly. This quick response ability has important survival value for us in the increasingly competitive world environment.

...The important thing to watch in solar applications is not how many systems are built *but who* builds or buys them. Our experience here in Santa Fe is that solar energy is attractive to young professionals in their 20's to 30's and to younger well-educated people in general,

FIGURE 73

especially those in the planning professions.

In Santa Fe many of us who are working in solar development feel we are participating in important work. We are writing history. In our search for simpler, more self sufficient and less expensive structures there is a real sense of dedication among those in the field.

...The question is: Are these people [developers *and* buyers] the lunatic fringe or are they the main stream of a changing American culture? It is my contention that the wild and wooly experimentors out in New Mexico, Northern California, New England, Manhattan, and other parts of the country, are that very important group of early innovators who are now adopting life styles and products that will later be used on a grander scale.

The one common element to many of the houses being built in Santa Fe is their relation to nature. The house is not just a fortress against the elements. It is designed to connect the occupant back into the natural processes for his own benefit. The house is alive. It functions. It changes and responds to the outside environment in a special way that supports and helps the people inside. You might say the home has a sort of consciousness. The structure is an organic protective skin that the owner wraps around his family.

In Santa Fe we are looking for a kind of primal structure. Someone is going to develop a simple low-cost home that gives the basic, primal functions of shelter and warmth. We will let you know what we find."

The preceeding three photographs show some of what Wayne has found.

Doug and Sara Balcomb

Dr. J. Douglas Balcomb is the same fellow who has published many important papers on passive solar utilization. He is also the Chairman of the New Mexico Solar Energy Association. The Balcombs decided that they wanted a solarium addition to their home in Los Alamos (not a prerequisite for the chairmanship). This unusual and pleasing design is the result (see photograph, next page).

The greenhouse is a 400 square foot addition to their frame home. Doug admits, "It's not really all for the plants. We plan to spend a lot of time out there." The sawtooth configuration on the roof and east side combined with the vertical glass in the south face guarantee no overheating problems in the summer. The west wall, except for the clear sawtooth panels in the roof, is opaque. Balcomb is concerned about summer light in the greenhouse and plans to paint all interior walls and the panelling in the second sawtooth white. The eastern wall is designed to allow morning light to flood through the exterior glass and into patio doors to the home.

FIGURE 74

The roof is supported by 4" x 12" hollow box beams that span twenty feet. There is a six inch drop from west to east for drainage. The entire solarium has a light open feeling like a high roofed cathedral, *with* pine trees growing through the roof.

Thermal storage is both passive and active. The passive is the earth in the floor and planting areas. The active system is interesting because it could very easily be added into any existing greenhouse. Hot air is tapped at the apex of the second sawtooth (the highest point of the greenhouse). The air is blown down ductwork (eight to 12 inch ducts would be about right for a small unit) into a rock-filled box. The box could also be filled with sealed water cans or jars. The hot air warms the thermal storage and the air flow to the greenhouse can be operated in several modes. (See Figure 75 on following page).

Doug told me, "I could have put the storage in the floor, I suppose, but this way we can use it as a planter box and grow things on top of it." This method appeals to me because it eliminates costly or exhausting excavation work. Plants love bottom heat and the top of the box should be one of the healthiest areas in the greenhouse. Also, should any problems develop in the storage or fans, there they are above grade, easy to get to and work on.

Dr. Balcomb estimates the finished greenhouse will cost about $2,500, and "more hours of my labor than I care to calculate." But look at the care and concern that went into the design, the two pine trees are saved! What a space!

FIGURE 75

Builder, Designer: Doub and Sara Balcomb

Cost: Approximately $2,500

Size: 400 square feet

Clear Area: Double Thermopane

Vents: 19 sq. ft. lower bottom of south wall; 27 sq. ft. upper in second sawtooth. Natural convection or reverse fan.

Heat to home: Natural convection through sliding patio door.

Storage: Above grade box, approximately 4' wide x 3' deep x 20' long, 240 cu. ft. rock filled, 1" insulation.

David Wright

David Wright is another New Mexico architect doing important work in passive solar design. The Clark Kimbell and Charlotte Stone home in Santa Fe includes a "green room" or plant area. The entire home, however, acts as a solar greenhouse, as it is designed to be a passive solar collector. The two-story south face is double glazed with commercial storm

FIGURE 76

doors. The continuous east/north/west wall is adobe, sheathed with styrofoam and stucco. The north corners of the house are rounded. Wright adopted this feature from a "D" shaped New Mexico pueblo (Pueblo Bonito). The configuration cuts down on north wind resistance, thus reducing heat loss.

Along with Wright's careful calculation and use of summer/winter sun angles, insulation and tightness are the most important solar features of this home. The roof and subfloor areas are heavily insulated. So is the front face, with an ingenious canvas-covered styrofoam shutter system. Wright has aptly dubbed this home the "adobe Thermos bottle." It is over 80 percent solar, with one wood stove supplying backup heat.

It is also a delightful living environment, offering all the creature comforts within an ecologically sound design.

Figure 78 shows another Wright designed home. This totally passive dwelling is owned by a Santa Fe builder, Karen Terry. It has an unusual orientation

FIGURE 77

FIGURE 78

for a solar building, a north-south axis. The home has been designed and built so tastefully that you really have to be looking for it to find it.

Herbert Wade

Herb Wade has designed a freestanding greenhouse with passive storage in the roof/back wall area of the structure (Figure 79). The 6 inch space between the rock storage and north wall creates a natural convection downdraft. The cool air falls down the wall and out onto the floor of the structure. The hot air is pulled through the front wire mesh into the rock and water storage. By including water drums in the rocks, Herb is combining the best features of both storage media.

There are about 22,000 lbs. of thermal storage, including the water, in this 12' long x 9' high x 8' wide greenhouse. That is a lot of weight but the triangular configuration of the storage bin, supported by posts and mesh, should keep it evenly distributed. The greenhouse is built on pilings and includes perimeter insulation and an anti-rodent screen.

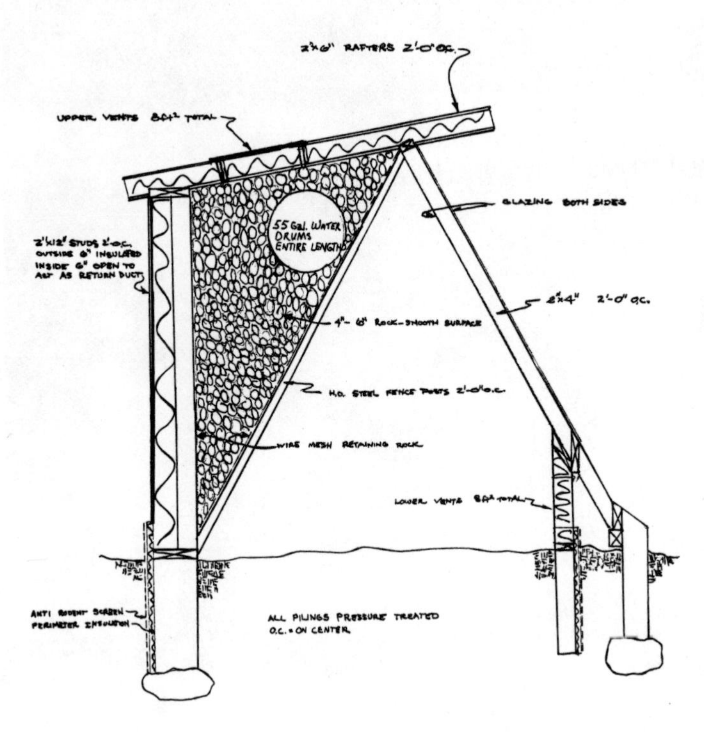

FIGURE 79

I think that this concept of integral wall storage could be applied to an attached greenhouse as well. Just allow for the openings to your house and tilt the roof southward to eliminate the roof valley between your home and the unit.

A Portable Homemade Greenhouse

So...you're a renter, or you move a lot. You want a greenhouse but don't want to leave it behind for the next occupants. Here's an alternative. Build a portable, but substantial, greenhouse that you can pack up and move to your next home.

This small lean-to was designed as a demonstration model to be moved around to fairs, energy exhibits, schools and the like. It's lightweight and can be assembled by two people. The largest panel is 8 feet x 8 feet and the panels fit into a rack on a pickup truck. Even though this is a display model, the greenhouse is fully functional, containing all the criteria for successful operation found in Chapter IV (insulated opaque walls, double skin,

FIGURE 80

shaded roof, etc.). When we set it up, we put in black water barrels and plants.

The five planes (you don't need a north wall, that's for display) are held together by slip pins in regular hinges. The intersections of the edges are sealed with foam insulating tape. The whole unit sets up in about 10 minutes with two people. Just set it down in front of your south window and you're in business. In a home application, it could be mounted on railroad ties (as explained in Chapter IV) right on the ground, or on a low block wall.

FIGURE 81

Designers: Paul Bunker, John Galt, Bill Yanda

Builders: Same as designers with help from the New Mexico Organic Growers Assn.

Size: 64 square feet

Cost: $256 (includes double fiberglass walls, north wall and No. 1 clear fir 2x2's for strength and uniformity). Funded by the New Mexico Energy Resources Board.

Clear Area: 125 square feet

Hardware: 2" sliding bolt hinges hold planes together; framing members pre-drilled then screwed together with No. 6 2½" wood screws.

Heating Capability for Home: In New Mexico, approximately 100,000 BTU's per day.

RESEARCH INSTITUTIONS AND ORGANIZATIONS

Sundwellings–Ghost Ranch Project–Abiquiu, New Mexico

The Sundwellings project is an important one. When the data are compiled and evaluated we will have an accurate accounting of the performance of various passive systems built in the same microclimatic area of identical materials. Perhaps more importantly, the project will demonstrate with hard data that low-cost energy systems built of indigenous materials are a viable alternative to plug-in high technological component systems.

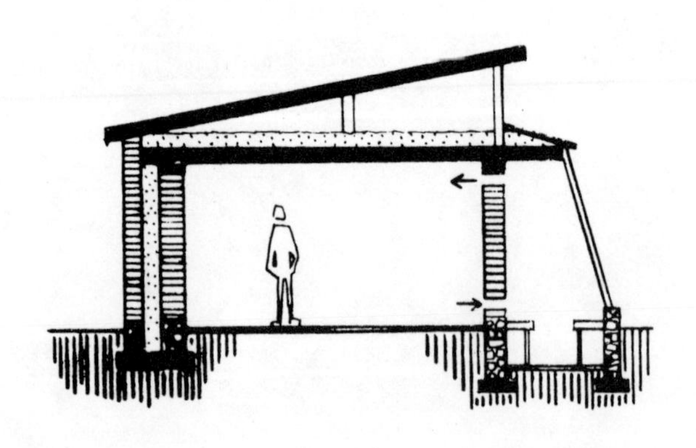

FIGURE 82

"Warm air from the greenhouse rises into the room while cool room air sinks into the greenhouse to be heated. Heat is stored overnight in the walls and floor of the room. The greenhouse is dug 2½ feet down into the ground to increase frost protection. The room can be closed off from the greenhouse to prevent unwanted heat loss or gain. The Northern New Mexico growing season can thus be extended from 4 months to 10 months."

United States Department of Agriculture

It's good to know that the resources of the American government are becoming committed to more energy efficient greenhouses. Robert C. Liu and Gerald E. Carlson designed the unit shown in Figure 83. An excerpt from their paper on the subject follows:

> A solar greenhouse must include a solar collector, a storage system, and a distributing system. It is our concept that fixed, south-facing collectors would occupy a position on the roof of the head-house, inside the greenhouse. or the north inside wall of the greenhouse. Placing the collectors inside the greenhouse should decrease heat loss from the collectors and may increase their efficiency. The protection to collectors that would be provided by the greenhouse roof, in addition to a possible increase in their efficiency inside the greenhouse, could lead to the development of a low-cost, unglazed collector that could be used year round for both cooling and heating. Actually, the whole greenhouse structure is a collector, and the excess trapped energy should be stored and reused.

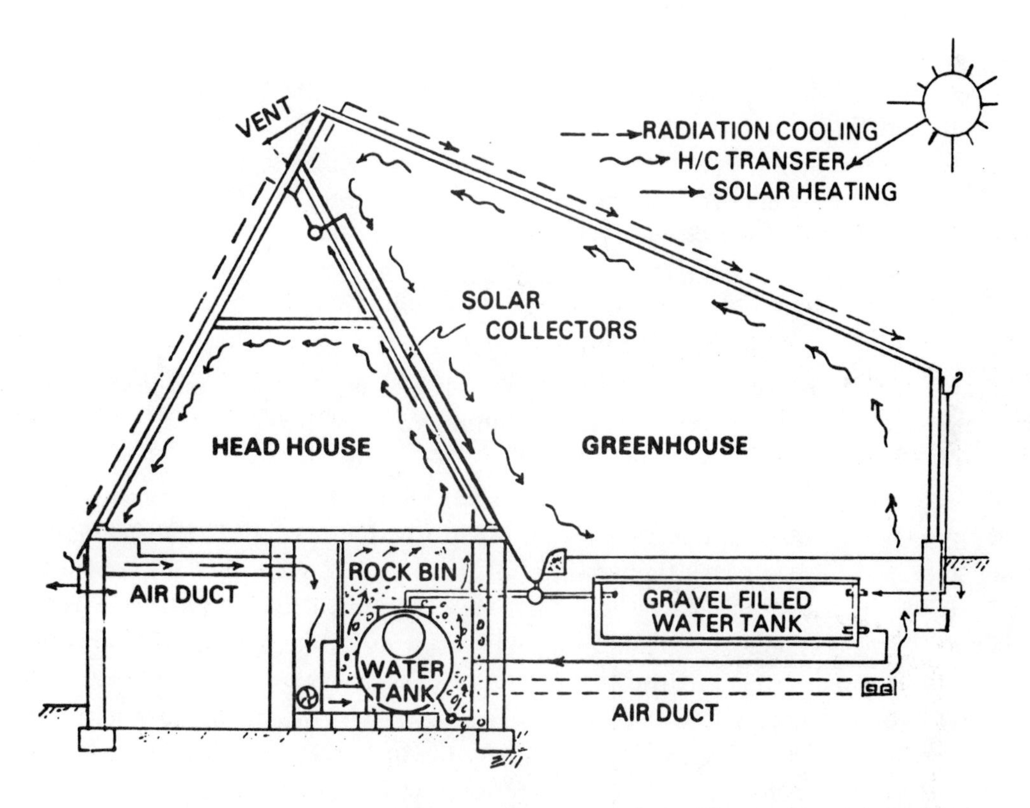

Sketch of a proposed new concept for maximizing solar energy use in a greenhouse.

FIGURE 83

The design utilizes both hot air and hot water. The hot water collector, which could consist simply of black polyethylene with a trickle and trough gutter below, is being tried by many of the larger research institutions.

This design is more complicated than the passive approach we advocate, but it is likely to be utilized in large, completely controlled greenhouse environments. A full report of Drs. Liu and Carlson's ongoing work can be obtained by writing to: United States Department of Agriculture, Agricultural Research Service, Northeastern Region, Beltsville, Maryland 20705.

The New Alchemy Institute

The New Alchemy Institute was formed in 1969 to confront future survival problems on our planet and to develop alternative technological solutions to them. The organization has concentrated its research on solar heated growing structures and the design of complex ecosystems. The institute is perhaps best known for its pioneering work in the field of aquaculture.

One project that New Alchemy is currently working on is the design and construction of a self-contained structure that will function as a food-and-energy-producing research center (see Figure 84). Located in Canada, the "Ark" will integrate fish raising in solar ponds with a large greenhouse running the length of the structure. The facility will be heated by the greenhouse and 800 square feet of solar collectors employing a rock storage system. Electrical power for the facility will be supplied by newly designed wind generators.

Another interesting design by John Todd of N.A.I. combines features of the pit greenhouse in a radical approach to solving the heat gain/loss problem. Todd designed it as a fish pond and Peter Van Dresser adapted it to a pit greenhouse. Only the south face would be clear, with the rest of the structure, including the roof, covered by an insulating earth berm. All interior surfaces would be *reflective*. The beauty of this design is that the major heat loss area (the entire south face) is one plane and could be tightly insulated at night. Since the south face would be covered during loss periods, it could be single glazed. Utilizing stabilized adobes, rough timbers, tin foil, and recycled glass, the unit could probably be built for 50 cents per square foot or less. The only possible drawback I see is that certain plants, such as tomatoes, cucumbers and beans, would block some light from the crucial side wall reflective surfaces. However, proper plant location and moveable interior reflectors might solve this problem. I think the idea has appeal, particularly in low light, extremely cold areas such as northern New England and the upper Great Lakes region.

The New Alchemy Institute is providing a vital service through its research into ecologically sound "solid states." N.A.I. is supported by private foundations and contributions; ($25 or more entitles you to their annual journal, a very worthwhile publication that reviews New Alchemy's continuing work. The journals can also be ordered from: Box 432, Woods Hole, Massachusetts 02543 (Volume 1 is $4; Volumes 2, or 3, $6).

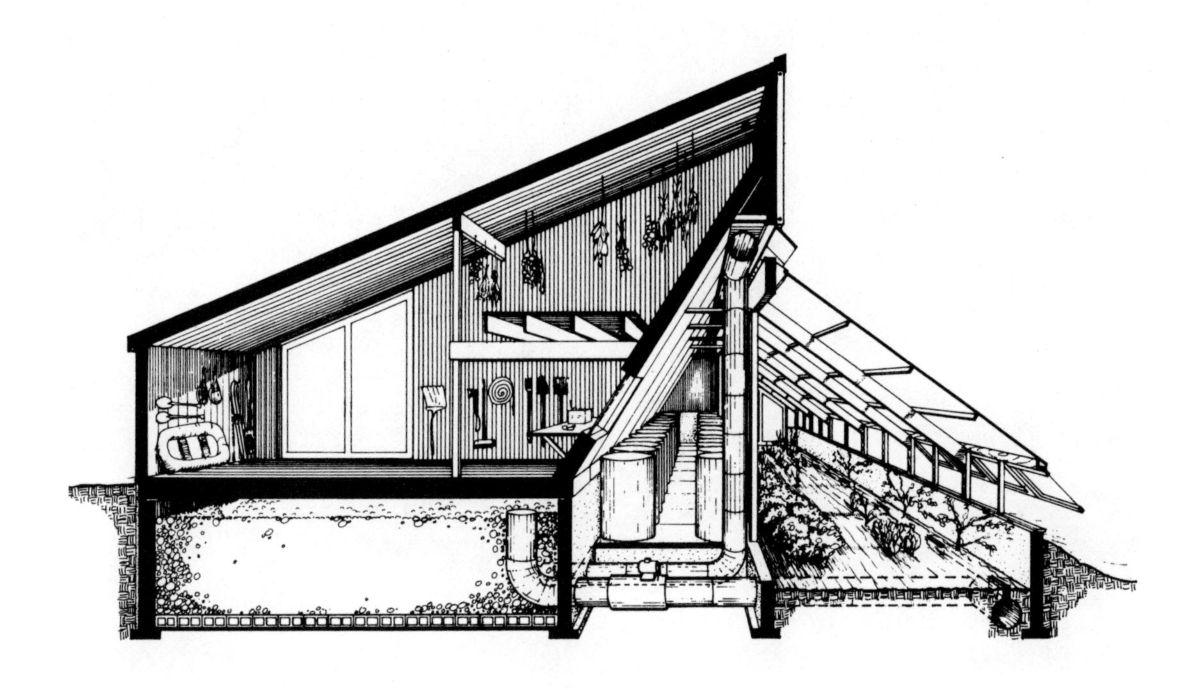

FIGURE 84

Greenhouse Heating and Storage

Solar Collection: 200 sq. ft. of greenhouse glazing exposed to southern sky.

Storage: 118 yard rock storage behind greenhouse and 19,000 gallon light transparent warm water fish culture facility.

Emergency Heat: Resistance-coils in air ducts.

Residential Heating and Storage

Solar Collection: 850 sq. ft., flat plate water, selective black.

Storage: 21,000 gallon tanks under living room.

Distribution: Fan coils

Supplemental: Woodstove and hydrowind power plant

University of Arizona Environmental Research Laboratory

This research center is concentrating on the development of integrated systems that provide power, water and food. Staff members at the laboratory are experimenting with aquaculture and vegetable production within polyethylene-covered greenhouses. Plastic-lined tanks contain shrimp that feed on organisms living among water hyacinths in the pond (Figure 85).

FIGURE 85

FIGURE 86

An interesting secondary project that is being developed is an insulating greenhouse cover similar to the Beadwall® that injects liquid foam between the plastic layers at night (see Figure 86).

Another facility created by the Environmental Research Laboratory expands the ecosystem approach to include desalination of sea water and nutrient preparation for the greenhouse use. This ambitious project is being carried out in collaboration with the University of Sonora and is located at Puerto Peñasco, Mexico (Figure 87). In this design, waste heat

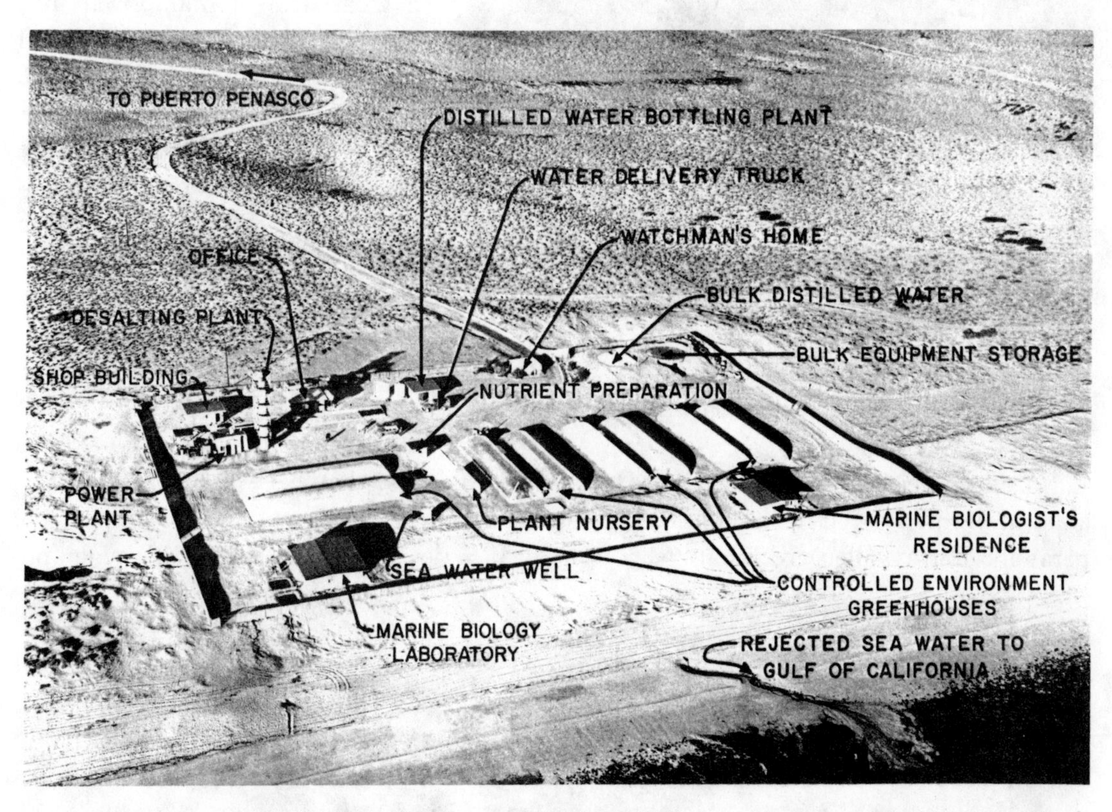

FIGURE 87

from engine-driven electric generators is used to desalt the sea water. The fresh water is then piped to vegetables within controlled-environment greenhouses of air-inflated plastic. The researchers state:

> The concept is applicable to vast regions where almost nothing grows—and where desalted water remains prohibitively expensive for open-field agriculture...The principal advantage of the concept for arid regions is extreme conservation of water. Moisture lost from field crops, by evaporation and transpiration, is enormous, of course. In a closed system this moisture can be captured. Estimates are that a plant within such a sealed-in environment uses only about a tenth as much water as it would need outdoors.

FIGURE 88

The long range potential for this kind of facility is fantastic. The Environmental Research Laboratory is presently directing construction of their first "large scale" installation in the Arabian Peninsula Sheikdom of Abu Dhabi.

Lately, the Lab has been putting a great deal of brainpower into thinking small. The experimental attached greenhouse shown here features their ClearView collector on the south vertical face and the liquid foam insulation in the arched inflatable roof. The ductwork shown in the photo is for demonstration and education and would not be visible in a residential application.

Bear Creek Thunder/Pragtree Farm/Ecotope Group

I am combining the work of these three groups because they have done a tremendous amount of interrelated research in the Pacific Northwest. For instance, take a look at the parabolic north walled greenhouse shown here. It is designed precisely for it's geograph-

FIGURE 89

ical area, a location noted for foggy and cloudy, but mild, winters. The solar heated/aquaculture greenhouse is located at Pragtree Farm (see Figure 90). Howard Reichmuth and Jeffrey Barnes did the engineering and design work on the structure. From Reichmuth's detailed paper on the description and analysis of the system we find:

FIGURE 90

> A passive solar heated aquaculture/greenhouse complex has been built 40 miles north of Seattle which collects and stores solar energy by reflection from interior surfaces into a massive interior thermal storage pool. Analysis indicates that such a scheme offers high collection efficiencies through low aperature temperature even for the case of relatively low quality reflectors with $\rho = .5$ (reflectivity of north wall).
> (In the region)...a solar energy collection system with a very low aperature temperature, approximately room temperature, can operate efficiently enough to provide up to 100% of the space heating needs. Ideally such a low temperature system should be passive, since an active system collecting at such low Δt would require high fluid pumping rates, consuming an unreasonable amount of energy for the energy collected.

Besides a comprehensive design analysis in the paper, I find some enlightening statements about the philosophy that guides Bear Creek Thunder's endeavors:

> The intention in this project has been to devise a long lasting building giving a premium to the simplest component choices, even if these component choices call for an adaptation in the use of the structure. Figuring out how to use the structure is a large part of the creative endeavor.

> A low storage temperature, high thermal mass, passive building using interior reflection is a productive approach to solar space heating in the Northwest. Such a structure can be built durably and economically

since the components of the collection system are also the building components and operate at low temperatures, avoiding degradation due to high temperature thermal cycling, as in a roof or a solar collector. Our experience of this building during construction suggests that solar heating which involves letting the light inside the structure is very dramatic and spiritually interesting. If we are entering a solar age, then it is appropriate that it actually feel like a solar age.

Howard told me that the only thing they've found so far that they might have done differently is not use the expensive R-20 insulation around the base of the structure and the fishtank. Evidently, the heat loss to the ground in this location doesn't justify its use.

Another design is the Rhombo-Cube Octahedron shown here. Ken Smith of Ecotope

FIGURE 91

writes: "It's a seasonal greenhouse (April-October) producing crops of Talapia grown from spawned fries. The greenhouse is fir pole covered with Monsanto 602, the tank a Sears 3000 gallon (12' diameter x 3' deep) vinyl pool liner. It contains 80 adult fish with an external filter system using a 1/12th horse power pump."

Yet another design being built is the solar greenhouse for a junior high school. The unique feature of this unit is the 200 eleven-inch cubic polyethylene clear plastic water bags stacked *directly behind* the south vertical window.

You can tell from this brief sampling the quantity, quality and direction of this work. They are nonprofit research groups and aren't exactly fat. If you write them for information, be sure to enclose a self addressed stamped envelope or you won't get an answer. Better yet, send them a donation and put your money to good use (that's my suggestion and was never mentioned by them).

Bear Creek Thunder, P.O. Box 799, Ashland, Oregon 97520
Ecotope Group, 747 16th East, Seattle, Washington 98112

Reichmuth's paper, "Description and Analysis of a Novel Passive Solar Heated Aquaculture/Greenhouse Complex near Seattle, Washington," is published in Volume 10 of the Sharing the Sun! proceedings, published by the American Section of the International Solar Energy Society.

Rural Research Housing Unit, United States Department of Agriculture, Clemson University

These folks are working on a three-year project to develop concepts, design, build and monitor residence/greenhouse combinations. They are funded by the Energy Research and Development Administration. The following is from the paper "Design Criteria for Greenhouse-Residences," by Harold F. Zornig, Martin Davis and T.E. Bond, from the proceedings of the Food and Fuel Conference (see Appendix C).

> The concepts shown were evaluated by a multidisciplinary team, including engineers, architects, and horticulturists. [The final design].... utilizes only the greenhouse as a solar collector and a rock storage system. The [first prototype]...is unique in that it has a roof-top solar collector serving both greenhouse and residence. Preheated air from the greenhouse passes through the roof-top solar collector before going into the underfloor rock storage system or directly into the house. In some areas, a greenhouse sized for the family's food needs may be too small to significantly reduce the heating load. With this design, the size of the roof-top collector can be adjusted to local heating requirements.
>
> During the summer, the greenhouse is shaded and vented to the outside for cooling. Convection air movement through the solar collector will draw air from the greenhouse at a rate of one-half air exchange per hour. This air is discharged through attic vents that are open only in the summer.
>
> The house will be cooled nocturnally during the summer by cooling the rock at night and circulating warm house air through the rock during the daytime. This should be an economical method of cooling in all locations where summer night temperatures average about 20 degrees lower than day temperatures.

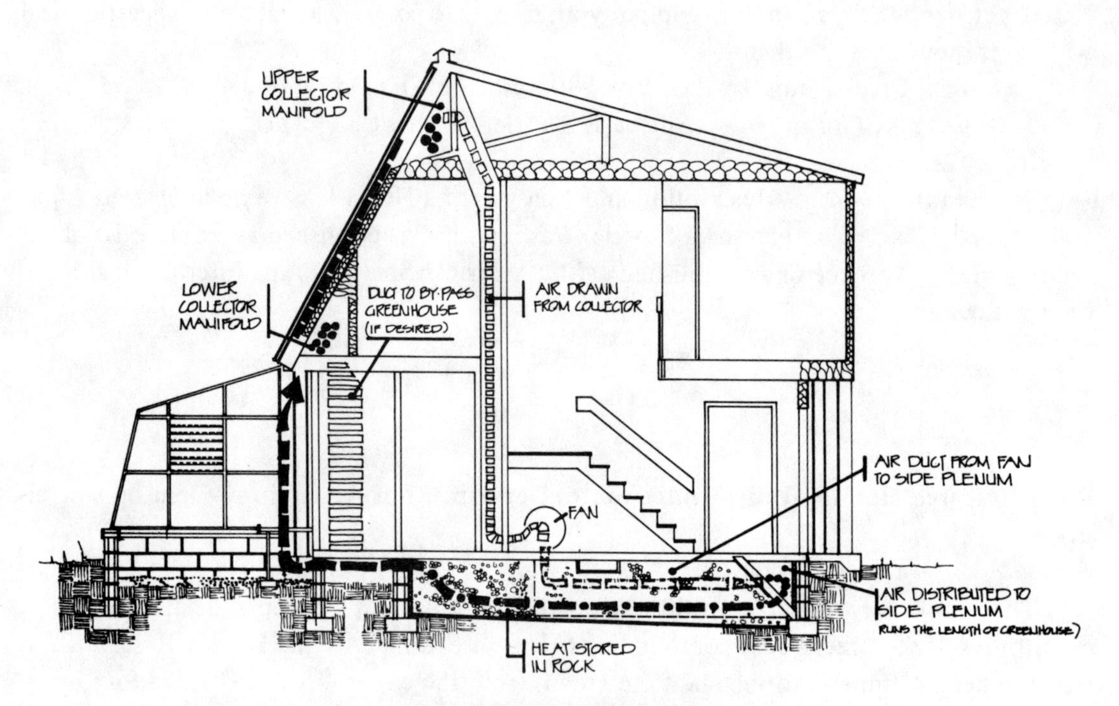

SCHEMATIC SECTION
COLLECTOR TO STORAGE CYCLE

FIGURE 92

Excerpts from the design criteria show:

Residence:
Three bedrooms, four occupants, 1,336 sq. ft. floor area, 10,688 $ft.^3$ heated space, occupied 50% of the time, one-half air exchange per hour.
Interior Temperature: Winter 65°F Summer 75°F
Interior relative humidity: Winter 60% Summer 70%
Average occupant CO_2 input: 2 ft^3/hr (total)
Average occupant O_2 demand: 3 ft^3/hr (total)

Solar Collector:
520 ft^2, average daily efficiency of 40% (Based on tests by Rural Housing Research Unit, Clemson, S.C.)
Average surface temperature: Winter 120°F, 9 hr., maximum 140°F
Summer 140°F, 13 hr., maximum 180°F
Designed rates of airflow through collector during winter operation are from 2 to 3 ft^3/min/ft^2 of surface.
Summer convection flow through solar air collection, with 8.5 $ft.^2$ opening and 120°F average temperature will be about 200 ft/min or 1,700 ft^3/min

Rock Storage:
1,200 ft^3 of clean railroad ballast (2 in. minus)
Weight: 94 lb/ft^3

Heat capacity: 19 BTU/ft^3/^{o}F
Static pressure of airflow through rock storage is 0.3 in. H_20 in prototype.
Average winter temperature: 82°F
Average summer temperature: 70°F
Average stored winter heat: 387,600 BTU (1¼ days storage @ 40°F outside)
Average stored summer cooling: 114,000 BTU (1 day cooling storage)
Heat loss to ground: 2,560 BTU/hr or 61,440 BTU /day (based on 4 BTU/hr/ft^2 loss) Rock storage tests by Rural Housing Research Unit, Clemson, S.C.

Greenhouse:
352 ft^2 of floor area
2,800 ft^3 of space
Heat loss by infiltration: Assume 10% in conduction loss
Ventilation for cooling: Winter–one-half air exchange per minute
Summer–one-half air exchange per minute
Shading: 50% in summer
Interior temperature: Winter–minimum 55°F, maximum 90°F (auxiliary cooling, cooled with water mist)
Summer–minimum 75°F, maximum 105°F (auxiliary cooling, cooled with water mist)
Relative humidity: January–average 70%, maximum 90%
July–average 70%, maximum 100%
Maximum hours of insolation: January–9, July–13
Insolation rate: Assume 60% absorption of available insolation
Note: Relative humidity in winter in both the greenhouse and residence will be controlled by condensation on the greenhouse single glass exterior. When the interior temperature is 9°F above the exterior temperature, a 90% relative humidity can be maintained. A 1°F rise in temperature is usually associated with about a 2% drop in relative humidity. Therefore, if the greenhouse is at 55°F and 90% relative humidity, the house at 65°F would be maintained at about 70% relative humidity, the average for January.
In the summer, relative humidity and temperature in the greenhouse will be controlled by ventilating all air directly outside through the solar collector. A ventilation rate of one-half air exchange per minute associated with 50% shading should maintain a maximum temperature of 5°F over the exterior temperature, or 105°F during the hottest (100°F) day. A water mist should be used for additional cooling.

The paper concludes by stating:

> Based on 1976 costs, families with annual incomes between $8,000 and $10,000 would save about $360 in food costs and $340 in heating costs. These families would also be helping to stretch the world's supply of fuel and food.

Zomeworks

Zomeworks and its president, Steve Baer, need no introduction as innovators in the field of solar energy. Three of this company's developments have important applications to solar greenhouses.

FIGURE 93

The "Beadwall"® system seen in the Monte Vista greenhouse was completed in 1973 (see Figure 93). It was invented by David Harrison of Zomeworks and solves the tricky problem of how to combine movable insulation and light-transmitting clear surfaces. In this system, insulating styrofoam beads are blown in between the two panes of glazing (3" apart) when temperatures drop below tolerable levels. In the morning, or when temperatures rise, a photocell switch tells the motors to evacuate the cavity and the beads are sucked back into their storage containers. I've seen Dave get a standing ovation from audiences at solar conferences when he was demonstrating this device. The Monte Vista greenhouse was one of the first applications of Beadwall® and used six vacuum cleaner motors and storage bins to accommodate the beads. Besides greenhouse applications, this system can be retrofitted or newly installed in home windows to maximize direct gain and almost eliminate heat losses.

"Nightwall"® is a poor man's Beadwall® in which *you* supply the motive power to move insulating panels. They are simply rigid styrene or styrofoam sheets that are cut to fit the exact dimensions of a window or greenhouse clear wall. Zomeworks supplies magnetic strips with sticky backs, small metal contacts to attract the magnets and an instruction sheet to help you put it together. When the magnets are attached to the perimeter of the window and the metal strips are adhered to the styrofoam panel, the magnetic force will tightly bond the insulation to the window at night. You remove the panels in the morning for transmission of light and direct gain. As the clear areas in a greenhouse are the major sources of heat loss in the structure, they are prime candidates for Nightwall® installation. The rigid panels could also do double duty as reflectors behind plant beds during the day. I tried a Nightwall® application in the Voute greenhouse. It didn't work too well but that was my fault, not the product's. I didn't put enough magnetic strips around the perimeter and I applied it to a nearly horizontal ceiling surface. Nightwall® will work well on a vertical or almost vertical surface in a greenhouse. Zomeworks also supplies a chart to show you how many BTU's the Nightwall® can save on standard sized windows in various parts of the country. It's amazing what these simple applications can do for you.

A third Zomeworks product with greenhouse potential is "Skylid"®. It's a device that uses counterbalanced weights and freon containers to open high vents or skylights. Steve uses Skylid in his home and greenhouse. Six skylids operate 240 square feet of insulated louvres to prevent heat loss. Baer's greenhouse also has six 30-gallon drums for thermal storage and 30 tons of rocks. This greenhouse is a single-glazed structure and has held 32-degree minimums in 5 degree outdoor lows. Baer is also using Skylid mechanisms as passive trackers for concentrating collectors. (Figure 94)

FIGURE 94

These active and passive improvements provide important options for the solar greenhouse owner. As we noted in Chapter V, the beginner might consider the simpler passive approach first. If you do want one of the more complex active systems, Zomeworks is definitely the place to go for it. They also sell plans for "bread box" water heaters. For more information, write to: Zomeworks, P.O. Box 712, Albuquerque, New Mexico 87103

Solar Technology Corporation

The Solar Technology Corporation of Denver, Colorado, is one of the few companies that presently markets a well-designed solar greenhouse. They offer three attached and free-

FIGURE 95

standing models. Deluxe Solar Garden (TM) features "enclosed thermal storage and a thermostatically controlled internal circulation fan..it provides for evaporative cooling and nighttime cool storage during the summer. Timer controlled automatic watering systems and a redwood decking complete the Deluxe Solar Garden."

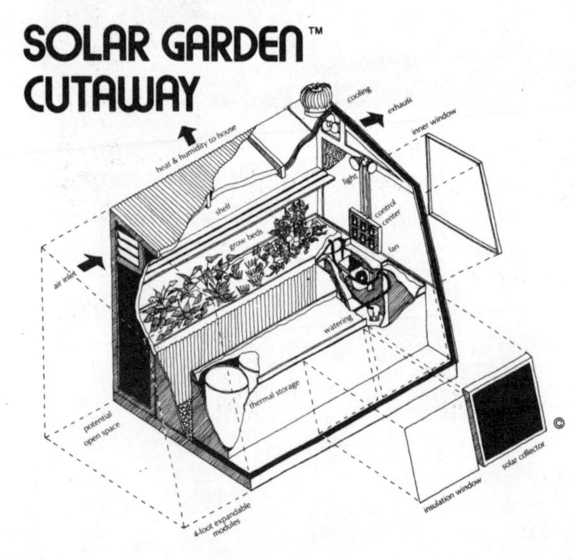

FIGURE 96

These greenhouses can eliminate the nightmares of the klutz builder. S.T.C. even offers solar consultants to analyze your building site, botanical specialists to supervise planting, and gardeners to help you maintain a healthy greenhouse environment. If this sounds good to you, a commercial unit may be exactly what you need. Just be sure that the design is appropriate for your area and conditions. For more information on the Solar Garden (TM) structures, write to: Solar Technology Corporation, 2160 Clay Street, Denver, Colorado 80211.

Helion

Helion is a group of young, but experienced innovators in multi-disciplines of alternate energy. The founder of the company, Jack Park, is assisted by Ken Johnson, a chemical engineer, and Richard Dehr. Their primary field is wind-powered generators and they have licensed the rights to a complete line of Kedco (an aircraft components company) high quality wind generators. Helion also manufactures a line of solar water heaters with accessories and an Energy Source Analyzer which gives a site survey with completely integrated outputs of wind and solar energy. Naturally, they needed a solar greenhouse to round out their efforts. The greenhouse pictured here is a first generation prototype. It was built over

FIGURE 97

Jack's swimming pool, which only goes to show the types of changes we all go through. The idea was to utilize the deep end for a cool air source as well as a root cellar. The frame over the pool was covered with plywood and tar papered for waterproofing purposes. The large front window faces true south and has an incline of 60°. The inner north side of the greenhouse is painted white to reflect light and all sides except the door are insulated with R-11 fiberglas batts. Only one 55 gallon drum was installed because of the weight factor on the light pool covering panels.

Ken Johnson's explanation follows:

> "The next step was to begin planting. Jack instigated this work in February starting with lettuce, tomatoes, cucumbers and cantaloupes. On the lettuce and tomatoes, an experiment was performed. In some cans of lettuce and tomatoes, soil was used as the growing medium. In other cans of the same vegetables, a hydroponic gravel was used as the growing medium. Horse manure leachate was used as a nutrient solution for all plants. During the course of the experiment no real advantage of using the hydroponic gravel was observed. However, significant heating and cooling cycles on the plants occurred during the experimentation period, causing severe damage to plant health. We can readily admit that this flaw in the system is due mainly to the fact that busy people sometimes forget to open and close vents. Work is being done to correct this problem. Ask any psychiatrist. Seriously, we do still have a ventilation problem and our next improvement to this greenhouse is to install a roof top wind turbine to see if this will help draw air through at a quicker rate. If this fails, our last resort will be to install a small bathroom fan at the top of the greenhouse inside, enclosed in a plenum chamber. During the day the fan would pull preheated air up to the top of the greenhouse, push it down between the double glazed windows into a bed of rocks, a heat sink, then exit out to the atmosphere."

In addition to those alterations, a shading lip protruding out about 18 inches over the top has been added on to help cool the unit. The main recommendation of Johnson and Park is to leave your swimming pool for swimming and put your greenhouse on solid ground.

With the brain power at Helion you can be sure that any manufactured greenhouse or plans would have the kinks ironed out. Write them for information: Helion, Box 4301, Sylmar, CA 91342.

The Vegetable Factory

The Vegetable Factory is in the business of manufacturing thermally efficient prefab greenhouses. They make both freestanding and lean-to models. You assemble the sections at home. From their literature..."Takes two 'unhandy' adults just four to six hours to assemble. Takes no experience. Only a screwdriver and drill are needed."..."For those desiring total window-clear transparency in specific areas, a double wall GE Lexan® panel is available."

The company has done extensive testing on their double fiberglass-acrylic walls. Their data indicate:

> Basically, glass, as a thermal barrier, is not very efficient. In actual practice, a single pane of 1/8" glass, the thickness typically used in greenhouses, conducts 36% more BTU's than a single pane of the glazing used in the Vegetable Factory panels. The Vegetable Factory wall panels are 244% more effective as insulators than single panel glass. They are 145% more efficient than double-insulating glass, the recognized standard in energy conserving glazing. The annual difference in fuel heating costs for a Vegetable Factory are reported to be 1/3 to 1/5 the cost of conventional single pane houses, relative to climatic area conditions and type of construction.
>
> Vegetable Factory's patented double-wall construction is 2 rigid panes permanently bonded on an aluminum I-beam grid for separation which creates a ½" dead air space between the layers of fiberglass. This substantially reduces the heat loss normally experienced in greenhouses.
>
> The double-wall fiberglass panels are very light in weight (each layer .025 weighs about .2 lb. per sq. ft.) eliminating the need for an elaborate, expensive understructure and foundation. Further, the large panel size (3' x 4' for roofs and sides), surrounded with elastic gasketry, eliminates most of the potentially expensive and leaky joints of typical greenhouses. Very little of the aluminum structure is exposed, preventing a great source of heat loss.

Their complete study of the transmission characteristics of fiberglass/acrylic is one of the best I have ever seen. The insulating percentages are higher than usual.

Vegetable Factory commissioned Garden Way Laboratories of Vermont to investigate the food production capabilities of the freestanding and lean-to model. I quote from their report:

> Winter bush varieties of squash are universally poor yielders, plus they cross-pollinate with summer squash and zucchini. The latter two should be the plants of choice.
> Carrots are too inexpensive to bother with.
> Forget about peas due to their rambling nature and the fact that their yield per foot of space is poor.
> Four hanging baskets of patio type tomatoes will yield over a ridiculously long period and are hardier than most standard varieties.

The photograph on the following page shows Garden Way's suggested block planting method. This method in a 5'6" W x 12' L lean-to model yields $245.64 worth of vegetables in a year at 1973 retail prices.

FIGURE 98

Lately, the Vegetable Factory has been exploring the solar heating potential of their lean-to models. They have an application on a private home in Nantucket, Massachusetts, based on the same principles as the lean-to designs in this book. For complete literature and pricing information, write Vegetable Factory, Inc., 100 Court Street, Copiague, Long Island, New York 11726

Solar Room

This is a young company with big plans. It is headed and founded by Stephen Kenin, the president of the Taos Solar Energy Association. Kenin is no newcomer to solar energy applications. He has worked for Zomeworks and with Hamilton Migel on his home (see page 94). Here is his explanation of the Solar Room:

> The Solar Room is a device that turns the southern side of a home into a solar heater. Made of a special plastic, a Solar Room can supply 35 to 65% of home space heating needs. With heat storage and insulation options its heating capacity is greatly increased. The Solar Room is available in kit form and is designed to be an exterior room, seven feet wide and as long as space permits; 20, 30 or 40 feet. The longer the Solar Room, the more heat is collected.
>
> Not only a heat collector, the Solar Room is a versatile, inexpensive addition to the home, costing only $2.50 to $3.50 per square foot of floor space. It is an airtight, thermally efficient space, and can serve as a greenhouse, a winter playroom for children or as a foyer to the house where coats, boots and bicycles can be stored out of the winter weather. As a greenhouse, the Solar Room is an especially efficient space, providing warmth for the household and fresh vegetables for the dinner table. In the Spring the garden can be started early in the greenhouse and transplanted outside when danger of frost is past.
>
> The Solar Room is also a "take-down" room. Because of its effectiveness as a heat collector it is not needed during warm weather, and has been designed to take down during the summer months. The initial installation requires less than a day's time, and after that removing the Solar Room in the Spring and putting it back again in the Fall takes only a few hours. When not in use, it takes up little storage space.
>
> The Solar Room kit is made possible by the use of an exotic new plastic that resists the disintegrating rays of the sun and thus lasts for years and years, special aluminum extrusions that hold the plastic in place, the best grade of clear heart redwood that will not rot in contact with the ground or in moisture, and galvanized ribs that support the plastic skin and will not rust. The Solar Room is double-glazed, which means there are two layers of plastic with an insulating "dead-air" space in between. It has withstood winds over 60 miles per hour.
>
> A 30 foot long unit costs about $500 and has a collector area of about 330 square feet, a heating potential that is conservatively in the millions of BTU's per heating season. It has been estimated by the Los Alamos Scientific Laboratory that a solar collector of 330 square feet may supply all of the heating needs of a 1,000 square foot house in this locale. However, in order to be conservative in the heat rating of our Solar Room, we claim that it will supply 50% of the heating needs of such a house. A Solar Room costs a fraction of the price of any

competing solar heating system of comparable heat collecting power.

The inflatable polygreenhouse, naturally, doesn't have the durability of a rigid fiberglass or glass house. However, it definitely has an initial cost edge over any similarly sized pre-fab solar collector. If the owner takes good care of the material and stores it in the summer it might go as long as five heating seasons before replacement of the inexpensive polyethylene. That's far beyond the pay-off period. As a matter of fact, in some applications like mobile homes (see photo below), the Solar Room could pay itself off in one heating period. (That's ignoring the food producing capabilities of the unit.) Any home owner getting hit for $150 to $200 a month heating bills (common in many areas) should certainly investigate this system.

FIGURE 99

Besides the inflatable Solar Room, the company is also developing water heaters and thermally designed window box greenhouses.

Write them for more information at Box 1377, Taos, New Mexico 87571.

APPENDIX A Sun Movement Charts

The charts can be used to determine:

1) The hours of direct sunlight your greenhouse will receive at any time of the year.
2) How obstructions will shade the unit.
3) The altitude of the sun at any moment.
4) A visualization of the sun's rays striking a flat or tilted plane.

You will need:

1) A small piece of graph paper.
2) A cheap protractor.
3) A straight edge.

Some definitions are in order:

1) **Sun Path**: The apparent (from our viewpoint) movement of the sun through the heavens. On our charts, the sweeping east-to-west lines are the sun's path on the 21st or 22nd of each month.

2) **Altitude**: The height in degrees of the sun *from* a true horizon. Altitudes are shown on the *concentric circles* at 9° intervals in the upper right of the charts.

3) **Azimuth**: The distance in degrees east or west of true south shown on the *radii* of the charts.

4) **Time-of-day**: The nearly vertical lines represent the solar (not time zone or daylight savings) times of day. They are noted across the top sun path line.

Here we go!

1) Find the chart nearest your latitude (Figures 100 and 101 are for 36° north latitude.)

2) On a piece of graph paper draw a scale model of the floor plan of your greenhouse. A size of *about* one half by one inch fits easily on the chart. Cut it out. You may want to cut out scale drawings of your home and any obstructions. To be accurate, all models must be measured and positioned to the same scale.

3) On the greenhouse model mark the junction of all solid and clear walls.

4) Place the model on the chart. By using the azimuth angles, position the model in its actual orientation. In our example, the site is facing true south so the model is parallel to the 90° east–90° west azimuth line. The greenhouse center should be in the exact center of the chart. If you have included a house, trees or other obstructions orient them

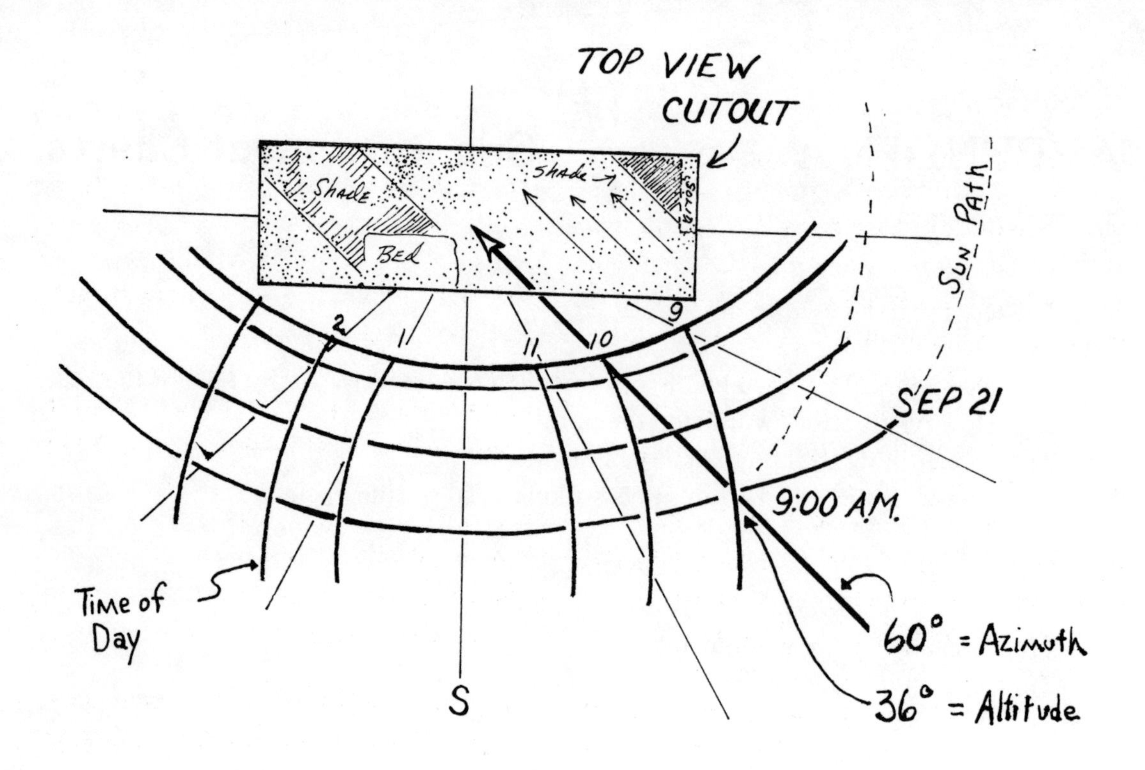

FIGURE 100

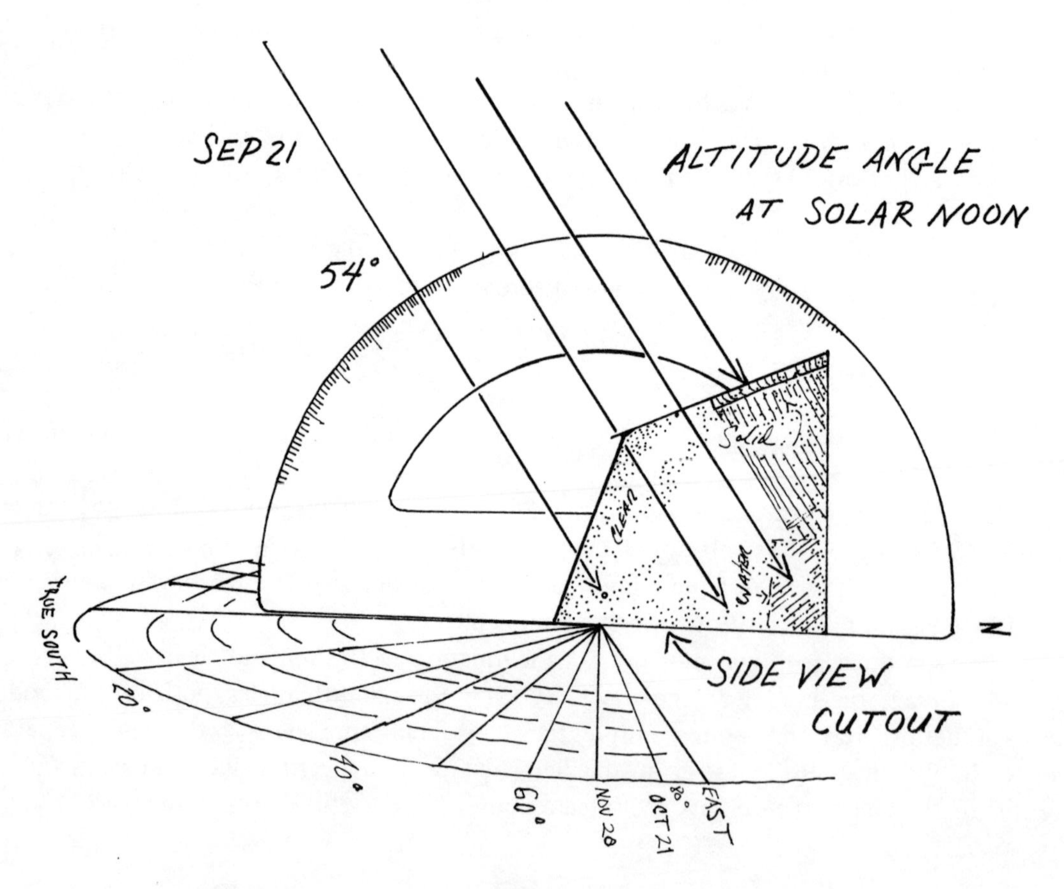

FIGURE 101

to scale and position them on the chart.

5) The first objective is to examine the duration of sunlight the greenhouse will receive at various times of year. Choose a month and follow its sun path, noting the number of hours the model is receiving direct sunlight. Solid walls and obstructions will shade the incoming light at certain times of day (see Figure 100). Observing the hours of shading will allow you to design clear and solid walls to best suit your particular location.

6) Next, to determine any solar altitude, find the point at which a chosen time-of-day line intersects the sun path line. Now find the nearest *concentric* circle and follow it around to the degree marking in the upper right. That's the sun's altitude. When you're in between circles–estimate.

7) To better visualize the angle of incoming light, place the flat edge of a protractor across the time-of-day/sun path intersection you are studying with the middle of the protractor over the center of the chart. Take a straight edge and connect the center of the protractor to the determined solar altitude angle on the edge. That's where the sun is at in that moment of time (see Figure 101).

8) A side view of the greenhouse cut out of paper is helpful in determining light patterns through clear roof areas and sides (Figure 101). By repositioning the solid/clear areas in the model, you should be able to get maximum winter sunlight for your location and also obtain some summer shading.

9) This same procedure can be used for any solar application.

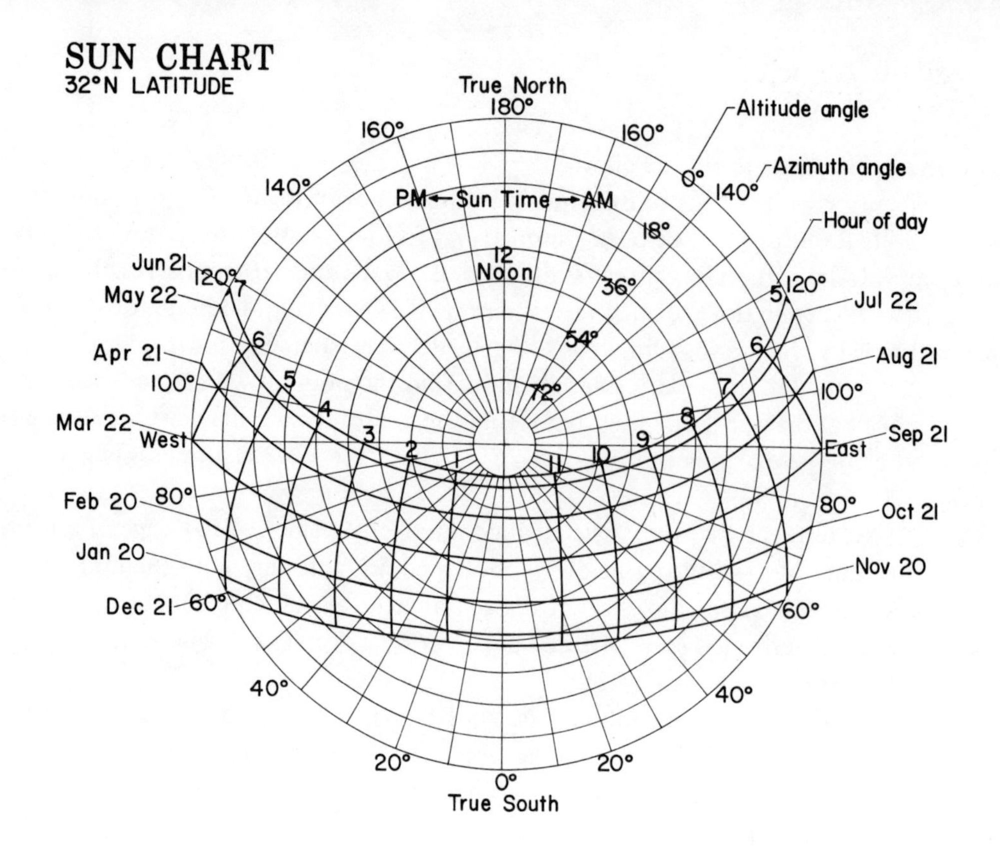

FIGURE 102

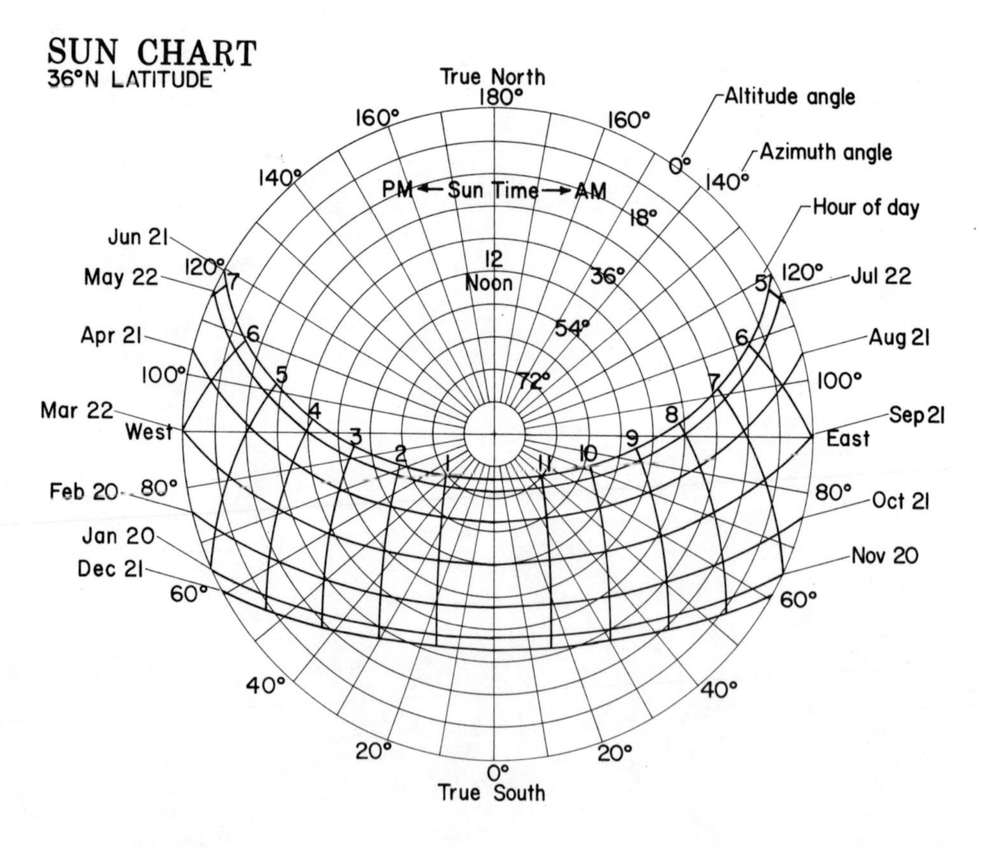

FIGURE 103

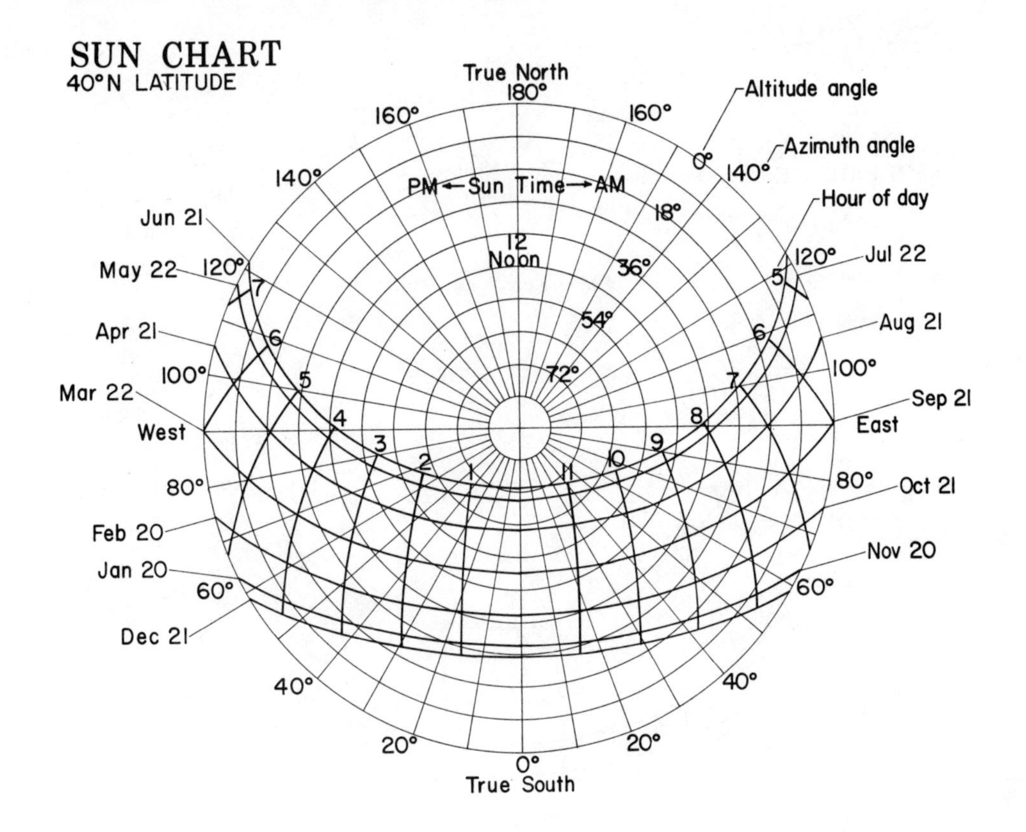

FIGURE 104

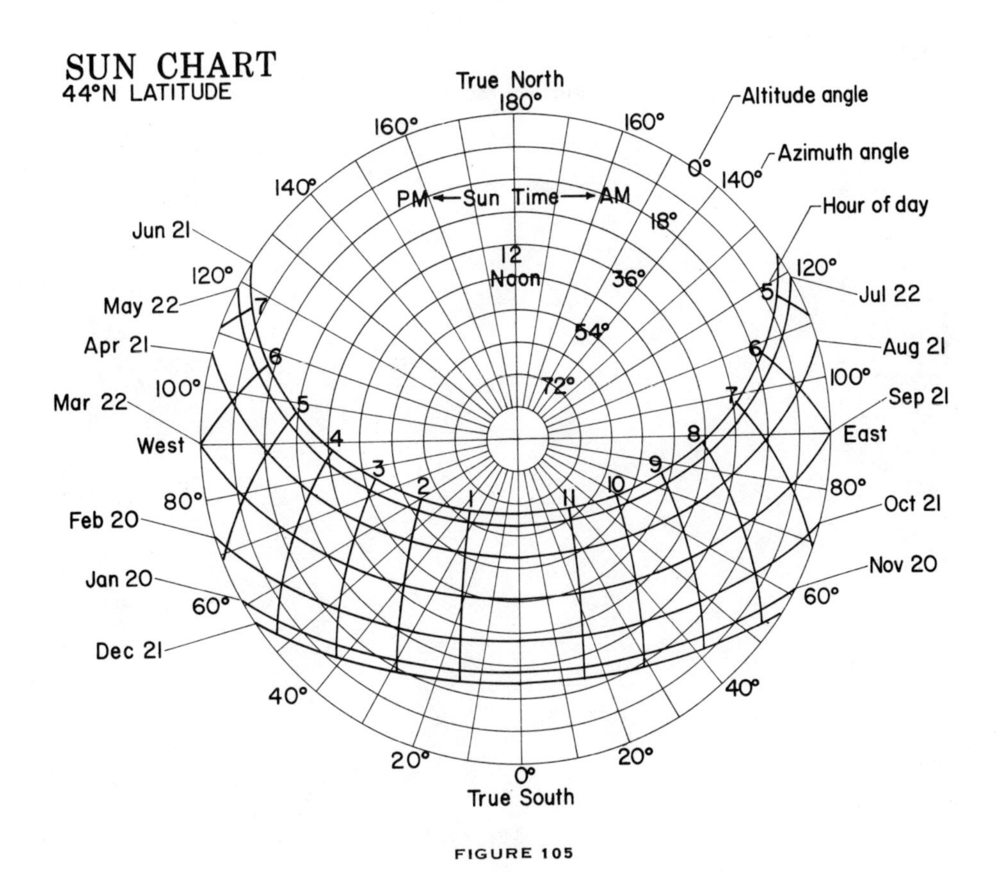

FIGURE 105

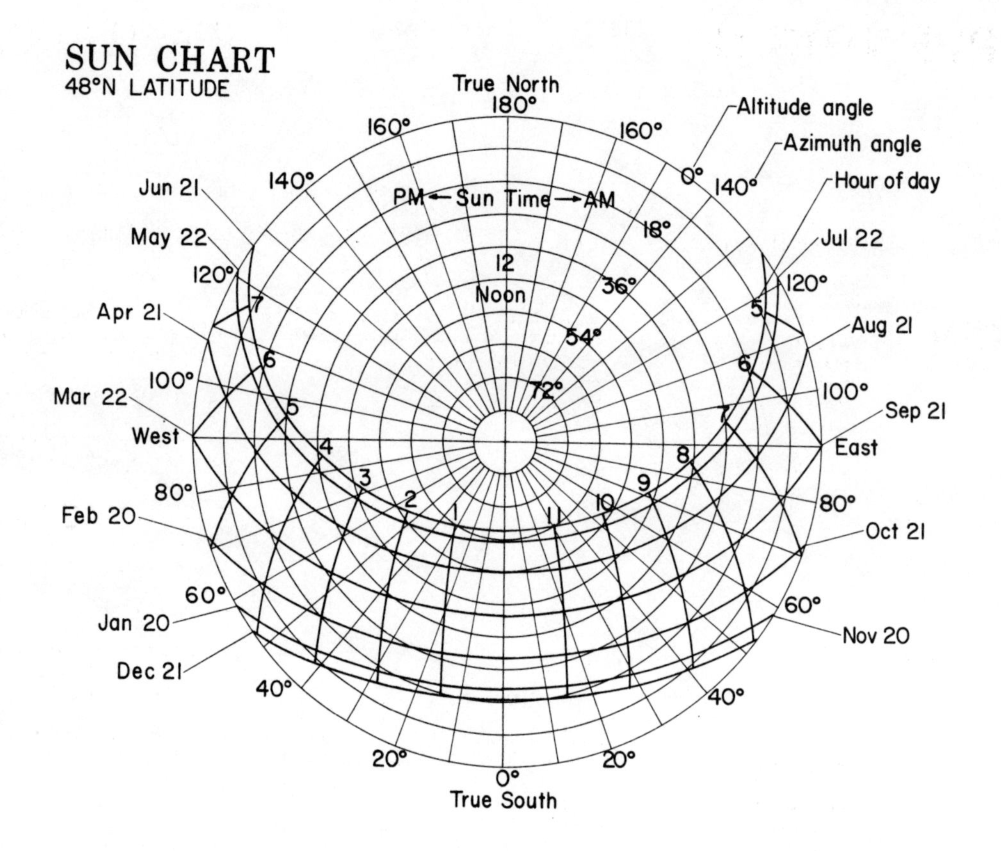
SUN CHART
48°N LATITUDE
True North
180°
160°
160°
Altitude angle
Azimuth angle
Hour of day
0°
140°
140°
PM ← Sun Time → AM
18°
12
Noon
36°
54°
72°
Jun 21
May 22
120°
Apr 21
100°
Mar 22
West
80°
Feb 20
60°
Jan 20
Dec 21
40°
20°
0°
True South
20°
40°
60°
Nov 20
Oct 21
80°
East
Sep 21
100°
Aug 21
120°
Jul 22
7
6
5
4
3
2
1
11
10
9
8
7
6
5

FIGURE 106

APPENDIX B Planting Chart

PLANT	VARIETIES	LOCATION	TIME OF YEAR TO PLANT	SPECIAL INSTRUCTIONS
TOMATO	EARLIANA BIG BOY MARGLOBE MICHIGAN OHIO EARLY GIRL ANY SMALL VARIETY PATIO, CHERRY, PEAR COLD SET FOR WINTER	FRONT OF GREENHOUSE IN SPRING. ON BACK WALL IN WINTER	EARLY SPRING MID-AUGUST	NEED FULL PHOTOPERIOD. POLLINATE BY LIGHTLY TAPPING OPEN BLOSSOMS OR SHAKE PLANT VIGOROUSLY. TRAIN PLANTS UP STRINGS. TRIM FOLIAGE SEVERELY WHEN INFESTED WITH INSECTS AND IN FALL TO PREVENT SHADING OF GREENHOUSE. DO NOT CUT TOP GROWTH UNTIL YOU ARE READY FOR PLANT TO STOP GROWING. TOMATOES ARE PERENNIALS AND WILL PRODUCE FOR A LONG TIME. PULL SUCKERS (FOUND IN CROTCH OF LIMBS) OFF. THEY CAN BE ROOTED: START IN SAND OR VERMICULITE.
CUCUMBER	ANY TYPE WILL DO, BUT EUROPEAN FORCING TYPES WHICH PRODUCE FRUIT WITHOUT ARTIFICIAL POLLINATION ARE EASIER.	CLEAR SIDE OF GREENHOUSE IN SPRING. BACK WALL IN WINTER. CAN STAND SOME SHADING.	EARLY SPRING MID-AUGUST	NEED FULL PHOTOPERIOD. POLLINATE WITH A SMALL BRUSH OR LET THE BEES IN. PULL OFF FIRST SEVERAL FEET OF BLOSSOMS FOR BETTER FRUIT SET. TRAIN ON STRING OR TWINE. CAN BE TRAINED TO CLIMB ALL OVER THE SIDES AND ROOF OF GREENHOUSE.
PEPPERS	RED CHILE ANY GREEN, BELL WAX	FULL LIGHT AREA	EARLY SPRING MID-AUGUST	ADAPT WELL TO SMALL CONTAINER OR BEDS. POLLINATE WITH SMALL BRUSH. BE CAREFUL NOT TO OVERWATER. CHILE PEPPERS DO NOT SEEM TO GET AS HOT AS THEY DO OUTDOORS; STILL DELICIOUS, HOWEVER.
MELONS AND SQUASH	WATERMELON CANTALOUPE PUMPKIN HONEYDEW CROOKNECK ZUCCHINI ACORN	NEED LIGHT AND LOTS OF ROOM. FRONT OF GREENHOUSE	EARLY SPRING	TRIM VEGETATION. GROW OUT THE VENTS--SUMMER. WILL CROSS POLLINATE: TRY TO SEPARATE VARIETIES BY DISTANCE.
LEAFY GREENS	LEAF LETTUCE ENDIVE KALE SPINACH MUSTARD GREENS CRESS CHARD COLLARDS CHICORY CELTUCE	MEDIUM LIGHT. COOL	ANY TIME--MAKES THE MOST SENSE IN LATE FALL, WINTER, EARLY SPRING	DEPENDABLE WINTER PRODUCERS. HEAD LETTUCE DOES NOT HEAD WELL IN GREENHOUSE. PLANT DENSELY, THIN AS YOU EAT. LEAFY GREENS WILL GROW IN POTS, ON VERTICAL SURFACES, ALMOST ANYWHERE. PLANT UNDER TRIMMED TOMATOES OR CUCUMBERS. CAN BE CUT MANY TIMES WHILE GROWING. WILL GO TO SEED IF TEMPERATURES GET TOO HOT.

PLANT	VARIETIES	LOCATION	TIME OF YEAR TO PLANT	SPECIAL INSTRUCTIONS
CARROTS	SMALLER VARIETIES	SUNNY	FALL WINTER EARLY SPRING	PLANT THICKLY AND THIN OUT. SLOW MATURERS. INTERPLANT WITH TOMATOES.
BEETS TURNIPS	ANY	MEDIUM LIGHT	FALL WINTER	DO WELL IN SHALLOW BOXES. PLANT THICKLY, BUT THIN TO ALLOW ROOT TO BECOME LARGE. FOLIAGE WHEN SMALL MAKES GOOD EDIBLES.
RADISHES	ANY	ANY PLACE	ANYTIME	DON'T PLANT MORE THAN YOU CAN EAT. EXCELLENT INDICATOR OF SOIL VIABILITY: SHOULD SPROUT IN 3 - 5 DAYS.
BROCCOLI CAULIFLOWER CABBAGE BRUSSEL SPROUTS	CALABRESE, ITALIAN SNOWBALL GOLDEN ACRE, CHINESE JADE CROSS	MEDIUM LIGHT. COOL	LATE SUMMER– EARLY FALL FOR WINTER HEADING	SPATIALLY CONSUMING CROPS. DO WELL IN POTS. TRANSPLANT WELL INTO GARDEN.
BEANS	POLE BURPEE GOLDEN BLUE LAKE	MEDIUM LIGHT. UP WALLS.	EARLY SPRING. LATE SUMMER.	GREAT ON NORTH WALL OF GREENHOUSE. TRAIN ON TRELLISES. CLIMBERS CAN BE USED ON GREENHOUSE EXTERIOR FOR SHADE (RED POLE BEANS).
EGGPLANT	BLACK BEAUTY EARLY BEAUTY	SUNNY	EARLY SPRING THROUGH SUMMER	POLLINATE WITH SMALL BRUSH OR FINGERTIP. TRANSPLANT FROM GARDEN BACK TO GREENHOUSE IN FALL.
PEAS	BURPEEANA EARLY BLUE BANTAM SNOW SUGAR	SHADY AREAS. COOL.	FALL WINTER EARLY SPRING	USED TO REPLENISH NITROGEN IN SOIL. FOUR POLES IN CORNERS, STRING LATTICES ACROSS.
ONIONS SCALLIONS GARLIC	ANY	MEDIUM LIGHT.	FALL WINTER	KEEP SOIL MOIST. FRESH TOPS ARE GREAT IN SALADS, TRIM REGULARLY. START SEEDS IN GREENHOUSE FOR GARDEN SETS. GARLIC IS AN INSECT FIGHTER, EITHER GROWING OR GROUND INTO WATER SOLUTION.
STRAWBERRIES	EVERBEARING	SHADE. UNDER TABLES	FALL	LIKE MORE WATER THAN VEGETABLES.
HERBS	WE HAVE HAD TREMENDOUS SUCCESS WITH ALL WE HAVE TRIED. THEY GROW IN ANY LOCATION, BUT PREFER MEDIUM LIGHT. MOST DO WELL IN COLD WEATHER.			

ANISE BORAGE CARAWAY CORIANDER OREGANO PARSLEY THYME DILL
BASIL CHIVES CHERVIL TARRAGON MINT (ALL KINDS) SAGE SWEET MARJORAM

TRANSPLANTS OUT

TOMATOES
PEPPERS
MELONS
BROCCOLI
EGGPLANT
SUNFLOWERS
CELERY
CUCUMBERS

BEST FOR TRANSPLANT FROM GREENHOUSE TO GARDEN. DON'T BE DECEIVED. START 6 - 8 WEEKS BEFORE ANTICIPATED LAST FROST. THEY GROW FAST. TRY NOT TO DISTURB ROOT SYSTEM ANY MORE THAN POSSIBLE: MELON AND CUCUMBER ROOTS ARE EASILY DAMAGED. HARDEN OFF BY EXPOSING TO OUTDOOR TEMPERATURES TWO WEEKS BEFORE TRANSPLANTING. GROW IN SMALL CONTAINERS WHEN YOU CAN: WE HAVE USED JIFFY POTS (THESE NEED TO BE CUT BEFORE BEING PUT INTO THE GROUND, THEY DO NOT DISINTEGRATE QUICKLY ENOUGH AND THE PLANT CAN BECOME ROOT BOUND), EXPANDING PEAT PELLETS, STYROFOAM CUPS, MILK CARTONS, MACDONALDS' QUARTER POUNDER

SQUASH CORN	CONTAINERS, TIN CANS. IF POSSIBLE, TRANSPLANT ON A CLOUDY DAY, SHADE AND WATER WELL AFTER TRANSPLANTING. DO NOT FORGET THAT ALL OF THESE SEEDLINGS PLUS FLOWER AND HERB STARTS ARE BIG SELLERS IN THE SPRING. A SMALL (160 SQUARE FOOT) SOLAR GREENHOUSE I BUILT IN IDAHO SOLD $273.00 WORTH OF SEEDLINGS ITS FIRST SPRING. THAT WAS OVER HALF THE COST OF THE GREENHOUSE.

TRANSPLANTS IN

TOMATOES PEPPERS EGGPLANT MELONS ONIONS BROCCOLI PETUNIAS PANSIES MARIGOLDS NASTURTIUMS GERANIUMS	IF YOU ARE CAREFUL, YOU CAN TRANSPLANT HEALTHY GARDEN CROPS BACK INTO THE GREENHOUSE IN THE FALL. BE SURE TO CHECK FOR INSECTS AND DISEASE FIRST. GET ALL THE ROOT SYSTEM YOU CAN, SHADE AND WATER THEM IN THEIR NEW HOME.

Notes

APPENDIX C Ongoing Research

The following list of papers indicates the amount and variety of work taking place in universities, institutions and private firms across the nation. Much of the research is aimed at solving problems for large commercial growers who make their living with greenhouses. However, the work is applicable to every greenhouse situation. Most of these papers are highly technical in nature and you'd be wasting time and resources ordering them unless you can understand and use the data.

I've tried to present the main emphasis of the papers by quoting from summaries or abstracts in order to give you an idea of what the paper's about. In some cases, we've used diagrams from the publication.

Many of the papers were presented at the Solar Energy—Fuel and Food Workshop presented by the Environmental Research Laboratory of the University of Arizona in cooperation with the Energy Research and Development Administration and the United States Department of Agriculture. I'm grateful to Merle H. Jensen for permission to quote from the proceedings of that conference. The entire proceedings are available for $5.00. Write to the University of Arizona, Tucson Environmental Research Labs, Tucson International Airport, Tucson, Arizona 85706. Papers from that conference are indicated by an asterisk. When there are many authors, I've listed the first author on the paper.

*C. Direlle Baird and David R. Mears, University of Florida. *Performance of a Hydronic Solar Greenhouse Heating System in Florida.*

> "A full-scale greenhouse located at Bradenton, Florida, has been equipped with a complete solar heating system. The system incorporates flat plate solar collectors, an external insulated storage tank, forced convection water-to-air heat exchangers, and automatic controls for both collection and greenhouse heating modes. Performance data on the various components and on the complete system are presented. From the summary...for systems following this pattern of operation, the design temperatures *must be low*, usually less than 100°F...however, larger storage systems and heat exchangers are required."

Drew A. Gillett. *A Paper on Solar Powered Greenhouses.* Kalwall Solar Components Division. Manchester, New Hampshire 03103.

The paper compares three small greenhouses:

1) A single glazed unit with very little thermal mass.
2) A single glazed unit with 1200 pounds of thermal mass.
3) A double glazed, north-insulated unit with 1200 pounds of thermal mass.

Tests were done in late May.

"...Greenhouse No. 3...had much higher minimum temperatures than either greenhouse No. 1 or No. 2. In fact, during the entire two week period monitored, its minimum temperature spread never fell below 55°F, considered acceptable for most plants nighttime temperature. Secondly, the average temperature spread for this greenhouse was on the order of 20 - 30°F which is also acceptable. The only time in which the thermal environment in greenhouse No. 3 was unacceptable to most plants was during a very sunny day following a previously sunny day. The fan and thermal mass system was unable to keep the temperatures below 92°F and in fact rose to 100° or so. Under these conditions it is felt that *some venting* must be incorporated to maintain reasonable temperatures in the greenhouse."

The paper also contains a design sketch of Greenhouse No. 3, temperature graph and data for test period, an economic comparison of different designs and an understandable sales pitch for Kalwall Solar Battery Collector Tubes, the water filled fiberglass tubes used in the experiment.

*John E. Groh. The University of Arizona. *Liquid Foam Insulation Systems for Greenhouses.*

"Liquid foam insulation is a method of reducing nighttime energy losses from roofs of greenhouses. This system utilizes a liquid-based foam insulation material placed at night between two layers of polyethylene in the roof of a greenhouse and has the calculated potential of reducing the night heat losses by 85%. Actual measurements on a test greenhouse have produced overall savings of 47% when compared to an unfoamed two-layered polyethylene greenhouse."

The system utilizes a foam generator to blow a soap bubble type foam between the air inflated polyethylene. (See Figure 86, page 122.) The foam is drained and recycled as the bubbles burst. So far, the system has been used only in areas where the nighttime lows do not get below freezing. This method has large-scale possibilities in the South in its present stage of development.

*Merle H. Jensen and Carl N. Hodges. The University of Arizona. *Residential Environmental Control Utilizing a Combined Solar Collector Greenhouse.*

This paper is an overview of a combination of projects at the Environmental Research Labs and how they apply to their home-greenhouse experiment.

"...The integration and performance of four concepts in environmental control are being evaluated: a venetian blind solar collector (see Peck, this section and photo on page 56, Chapter V), rock storage for heated or cooled air, advanced methods of evaporative cooling and liquid foaming of a greenhouse ceiling for nighttime insulation."

Of particular interest to me is a section on the vegetable production capabilities of the unit and planting by "light requirements" of the crop.

Table 1–Selected Cultivars

Vegetables	Cultivar	Plant or Seed Source
Pepper	Christmas Pepper New Ace	George Ball Pacific, Inc. Takii and Co., Ltd.
Lettuce	Dutch type Summer Bibb Salad Bowl	Rijk Zwann Joseph Harris Co. Joseph Harris Co.
Tomato	Patio Tiny Tim Small Fry	George Ball Pacific, Inc. George Ball Pacific, Inc. George Ball Pacific, Inc.
Eggplant	Beauty Hybrid Black Prince	George Ball Pacific, Inc. Takii and Co., Ltd.
Cucumber	Patio Pik	George Ball Pacific, Inc.
Spinach	Malabar	W. Atlee Burpee Seed Growers Co.
Radish	Champion or Red Prince	Joseph Harris Co.
Fruits		
Strawberry	Fresno	California Strawberry Foundation
Banana	Dwarf	Environmental Research Laboratory
Herbs		
Annuals and Perennials	Assorted	George Ball Pacific, Inc.

George Ball Pacific, Inc., Box 9055, Sunnyvale, CA 94088

Takii and Co., Ltd., Umekoji-Inokuma, P.O. Box 3091 (Kyoto Central) Kyoto, Japan

Rijk Zwaan's Zaadteelt en Zaadhandel B.V., Gurgem. Crezeelan 40 De Lier, Holland

Joseph Harris Co., Moreton Farm, 3670 Buffalo Rd., Rochester, NY 14624

W. Atlee Burpee Seed Growers Co., P.O. Box 6929, Philadelphia, PA 19132

California Strawberry Foundation, Plants Nursery, 3570 Old Alturas Road, Redding, CA 90001

*A. W. Gerhart. Gerhart & Son Greenhouse, Inc., North Ridgeville, Ohio. *Practical Application of Energy Saving Ideas.*

Mr. Gerhart is a commercial greenhouse vegetable grower in Ohio and his short paper gives some excellent data on costs, problems and practical applications his company has experienced. Here are some excerpts:

> "...we require two and a half to three people per acre to grow a 200,000 pound tomato crop...we have to spend $24,000 per acre for natural gas...$30,000 to $34,000 for No. 2 heating oil...and an estimated $18,000 for coal (1975 prices)...the trick is to burn a cheap coal efficiently and still satisfy the Environmental Protection Agency.
>
> ...when the price of oil went up four times, the price of greenhouse tomatoes did not increase four times.
>
> ...we have adopted...the drastic reduction of fresh air into our plant. We simply do not introduce any fresh air into the greenhouse unless it is necessary to cool the crop. We know the plants need carbon dioxide (CO_2) and that many kinds of fungus and bacteria thrive in high humidities. We introduce CO_2 constantly, starting an hour or so before sunup and continuing until two hours before sunset. Our CO_2 generator is water cooled, so most of the moisture is removed that would be a part of the combustion process. This heat is then released into the greenhouse through a heat exchanger or, if the sun is shining, the heat is stored in our 100,000-gallon cistern until nighttime and then introduced to the greenhouse. Our irrigation water comes from the cistern and generally we irrigate with water of about 80 to 90°F. We use fungicides to control botrytis and leaf molds, a procedure we have used successfully the past four years.
>
> The next item of interest is the installation of aluminum foil behind our steam heat pipes to reflect heat back into the greenhouse. We use building foil with paper on one side and discard it after the winter season.
>
> For years we have lined our outside walls with plastic sheeting, but this year we installed a 4-mil double wall cover over our glass, and inflated the cover as is done in plastic houses. There will certainly be more potential in the future for us to use the double wall system.
>
> Inside we have been experimenting with methods to improve production without increasing energy consumption. One procedure that seems to be effective is to warm the soil to approximately 70°F. We have lowered our air temperatures from two to four degrees, and we are still getting excellent production.
>
> Perhaps the most intriguing energy utilization procedure with which we are now working is the use of a reflective material on the ground beds to reflect solar energy back up to the plants. We first started this

> when a representative of a firm approached us with the idea of using this material as a heat reflector over the top of the crop at night. We were not enthused about this procedure, but offered to test the material as a light reflector on the soil. The first year we showed a ten percent increase in production on a small plot. This year we have one fourth of an acre covered with Foylon (manufactured by Duracote), and to date we have an increase of 15 percent over the check areas.
>
> Also, we use artificial lighting in growing cucumber transplants. Florescent tubes are mounted five to six inches from the plants, with the lights on 24 hours a day. Two weeks from the seeding date, the transplants are ready for ground beds. We use 100 square feet of area to grow 12,000 plants. The excess heat from the light is used to help heat our service room."

At the end of the paper, Mr. Gerhart expresses optimism about his industry being able to survive the energy cost increases. The entire paper should be required reading for anyone thinking about entering the commercial greenhouse food growing business.

Lawand, et. al. Brace Research Institute of McGill University, Quebec, Canada. Two papers—*An Investigation of the Contribution of Solar Energy in Heating Greenhouses in Quebec* and *The Development and Testing of an Environmentally Designed Greenhouse for Colder Regions.*

It was the Brace Institute people who first built and evaluated an insulated and reflective north wall greenhouse. (See The Herb Shop, page 87.) They can also claim a great deal of responsibility for awakening the scientific community to the idea of the greenhouse as a collector. The first paper is an engineering analysis of the total solar energy a test greenhouse in Quebec received in a measured winter period. The second is on the evolution of the tilted north wall greenhouse. They also distribute inexpensive plans for their greenhouse and other low cost owner built solar energy devices.

*David R. Mears and C. Direlle Baird. University of Florida. *Development of a Low-Cost Solar Heating System for Greenhouses.*

> "A system has been developed and tested for storing large quantities of warm water under a bench in a greenhouse and transferring the stored heat to the greenhouse when needed. The unit can be constructed of low-cost material and therefore can have large storage and heat transfer capacity.
>
> Relatively low-cost polyethylene film solar water heaters have been built and their performance characteristics determined. These collectors would appear to be more economical to operate than conventional flat-plate collectors at relatively low temperatures."

In this paper the test results on several types of collectors made with inexpensive materials and assorted configurations of black poly absorbers, mesh shadecloth and clear poly coverings are examined. Flow rates and efficiencies of heating and losses are determined. In the system, the warm water is then drained off the bottom of the collector and stored in an insulated tank that fits under the greenhouse benches. The warm water in the tank heats air moving through a poly tube which in turn heats the greenhouse. See Figure 107

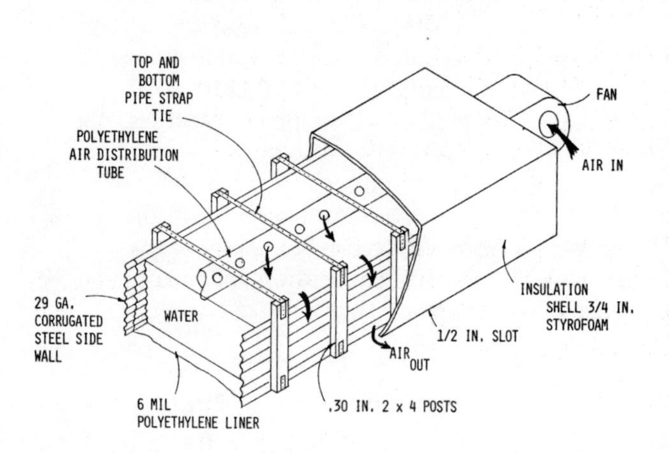

FIGURE 107

David R. Mears, et. al. Cook College, Rutgers, New Brunswick, New Jersey. *New Concepts in Greenhouse Heating.* Published by the American Society of Agricultural Engineers, St. Joseph, Michigan.

> "Heat loss due to longwave infrared radiation from double filmed, air-inflated polyethylene greenhouses can be significantly reduced. Methods of transferring low quality heat from warm water to air are proposed. Their feasibility is enhanced when radiation losses are reduced. Industrial waste heat and solar collectors are considered as warm water sources."

Robert T. Nash and John W. Williamson. Vanderbilt University, Nashville, Tennessee. *Temperature Stabilization in Greenhouses.*

This paper is a technical examination of the structural heat loss coefficient of a test greenhouse, the thermal storage coefficient and the heat transfer coefficient between the greenhouse air and storage system. (In this case, water in drums). It also contains some interesting observations from other sources:

> "...Many plants require a variation of the air temperature for optimum growth. Went has established that there will be a best nyctotemperature, or nighttime temperature and a best phototemperature, or day time temperature for any plant...Wellensiek stated...It has happened in Holland that very ambitious tomato growers have gotten up in the middle of the night to heat their furnaces, but nevertheless had poorer crops than their lazy neighbors, who let the night temperature fall."

A continuation of Prof. Nash's work with thermal storage is found in *Thermally Self*

Sufficient Greenhouses published by the International Solar Energy Society in Volume 7 of the Sharing the Sun! proceedings.

*John F. Peck, University of Arizona. *Basic Solar Collector Design and Considerations.*

> "This paper presents a few basic principles of collector design and use...It also presents a framework with which to organize the various general types of solar collectors; that is, a system of solar collector 'taxonomy'."

There are some easy-to-understand graphics in this paper as well as an explanation of the University of Arizona's "Clear-View" collector. This venetian blind type collector can adapt to a variety of operating modes and offers good potential for greenhouse development (see Franta, page 103 and Improvements, page 56).

*D.R. Price, et. al. Cornell University. *Solar Heating of Greenhouses in the Northeast.*
This paper covers a wide range of topics. To quote from the abstract:

> "...To reduce heat loss at night from a greenhouse, the researchers have investigated various insulation techniques for use at night that may be removed during the day. *Pulling black cloth* along the walls and across the ceiling reduced nighttime heat loss by approximately 50%.
>
> ...An analysis of using only the greenhouse as a collector versus using a combination of the greenhouse and external collectors is presented. From a cost effectiveness standpoint, there is justification for *using only the greenhouse* as a collector with modifications to improve its efficiency." (Remember, this is upstate New York, *not* sunny New Mexico...B.Y.)

There is also a listing of advantages and disadvantages of using the greenhouse as a collector compared to adding an external solar collector in the paper.

*W.J. Roberts, et. al. Rutgers University. *Using Solar Energy to Heat Plastic Film Greenhouses.*

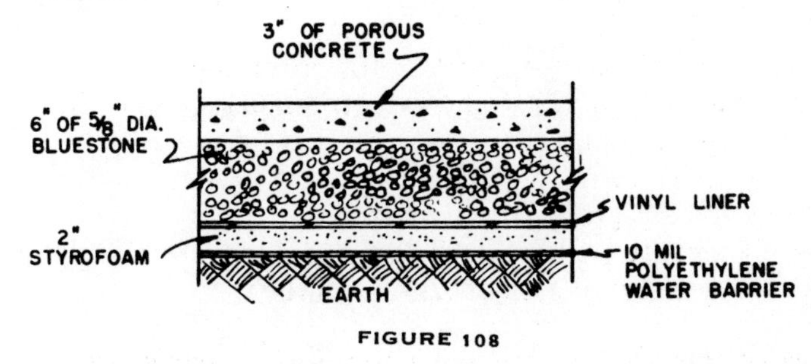

FIGURE 108

The paper has a section on black plastic water flow collectors sandwiched between two air inflated sections of clear film. It also contains data and diagrams of a porous concrete floor which acts as storage for the solar heated water (see Figure 108). The collection and storage system are linked together with an air inflated greenhouse that utilizes a moveable black plastic insulating curtain.

Mr. Roberts is an acknowledged expert in calculating the heating and venting needs for greenhouses and has published a good booklet on those requirements entitled *Heating and Ventilating Greenhouses*, Cooperative Extension Service, Cook College, Rutgers–The State University of New Jersey, New Brunswick, New Jersey (in cooperation with the U.S.D.A.)

Jack Ruttle, *A Nice Greenhouse for Nasty Winters: All Solar.* Organic Gardening and Farming Magazine, September 1976, page 69.

This is a non-technical presentation of Dave MacKinnon's work (see page 96) with a free-standing solar greenhouse at Organic Gardening and Farming's New Farm. The unit maximizes insulation on the north, east and west walls, the walls losing only 500 BTU's per hour at a 30° temperature difference with the outside. The design minimizes the clear areas. It also mentions Dave's designs of removeable polyethylene panels to cut nighttime heat losses through the clear surfaces. The relationship of low nighttime temperatures (relative to conventional greenhouse management techniques) was touched upon.

> "Bob Hofstetter...claimed superior plant growth in the solar greenhouse. Now he feels that constant warm temperatures *encourage* weak, spindly growth–and that the cool nights held insect pests to a minimum. The solar greenhouse produced excellent lettuce, excellent chard and even some early spring tomatoes!"

I'd certainly like to see more hard research done in the area of temperature vs. health and plant growth and increased research in cold and disease tolerant crops. It's easier and cheaper to plant the right seed than gain 10° in January.

Joel C. Simplins, et. al. Rutgers University, New Brunswick, New Jersey. *Reducing Heat Losses in Polyethylene Covered Greenhouses.* American Society of Agricultural Engineers, St. Joseph, Michigan 49085.

Extensive hard data on the thermal characteristics of various films and curtains used for exterior and interior greenhouse coverings.

*Ted H. Short, et. al. Ohio Agricultural Research and Development Center and the Ohio State University. *A Solar Pond for Heating Greenhouses and Rural Residences–A Preliminary Report.*

The paper explains the problems and potential of utilizing an independent solar pond for heating applications.

> "The solar pond is heated by solar radiation passing through the salt

water to the black liner holding the liquid. As the black liner temperature increases, heat is transferred to the 20 percent brine in the bottom half of the pond. The heated 20 percent brine rises no higher than the bottom layer of the gradient, and cooler 20 percent brine moves down to replace it. The upper non-convective region is nearly transparent to incoming ultraviolet and visible radiation, and nearly opaque to incoming infrared and outgoing long-wave re-radiation. One meter (39.5 inches) of non-convective water is a good insulator with a conductivity equivalent to approximately 6 cm (2.36 inches) of styrofoam. Since the walls are also insulated, losses are reduced significantly.

A major advantage of the solar pond is that both summer and winter radiation can be collected and stored for later use. After a full summer's radiation, the pond temperature throughout the bottom half should approach boiling. The OARDC pond will be limited to 80° C (180°F), however, to maintain liner stability. This upper temperature limit will be controlled with discharge heat exchangers or by covering the pond with an opaque plastic film."

Greg Stone, *Greenhouse Coverings: The Choice is Yours.* Organic Gardening and Farming Magazine. September 1976, page 58.

An informative, non-technical discussion of the aesthetic, economic and utilitarian factors involved in choosing the clear covering for your greenhouse. Good overall presentation of the multitude of options available.

*J.W. White, et. al. The Pennsylvania State University. *Energy Conservation Systems for Greenhouses.*

Testing and evaluation of thermal blankets and the problems involved in installation and maintenance are discussed in this paper. Also, some "significant generalizations" concerning the use of phase change solution such as salt hydrates and aqueous eutectic are presented. The Penn State research concluded that "...the utilization of conventionally-constructed lapped glass greenhouses as solar collectors does not show potential in Central Pennsylvania between November 15 and February 15." (If this seems to be in contradiction to the D.R. Price paper on New York greenhouses as collectors, you'll have to check out the parameters for yourself to decide.) The paper also includes interesting sections on utilizing stagnation in greenhouses and a self-fogging roof.

*James B. Wiegand. President, Solar Energy Research Corporation, Granite Building, 1228 15th Street, Denver, Colorado 80202. *Greenhouse Solar Heating: Techniques and Economics.*

"The modes of operation and performance testing done at Solar Gardens, Longmont, Colorado, are evaluated. Alternate schemes and configurations of solar heating in greenhouse structures are discussed and a cashflow analysis for a trial project is exhibited. Products useful in greenhouse solar heating, including controls available now, and collector and heat storage systems available soon, are discussed briefly. Practical considerations in adapting existing greenhouse structures to solar heat are discussed."

Mr. Wiegand provides some honest pros and cons for applying solar technology to existing buildings. However, I'm skeptical about his 20 year "cashflow analysis" on a solar greenhouse...I occasionally have to do those things myself.

BIBLIOGRAPHY

Chapter II The Dependence Cycle

Small is Beautiful, E.F. Schumacher, Harper and Row, 1973
Radical Agriculture, Richard Merrill, editor, Harper and Row, 1976
The Survival Greenhouse, Jim DeKorne, The Walden Foundation, 1975, El Rito, NM
Sunspots, Steve Baer, Biotechnic Press, 1975
A Landscape for Humans, Peter van Dresser, Biotechnic Press, 1972
The Daily Planet Almanac, Pacific High School, John Muir Publications, 1976

Periodicals and Papers:

The Coevolution Quarterly published by The Whole Earth Catalogue, edited by Stewart Brand and J.D. Smith, published quaterly.

The Mother Earth News published bi-monthly at Hendersonville, N.C.

Organic Gardening and Farming published monthly by Rodale Press, Emmaus, Pa.

Chapter III The Design

Direct Use of the Sun's Energy, Farrington Daniels, Ballantine Books, 1964
The Coming Age of Solar Energy, D.S. Halacy, Harper and Row, 1973
Energy Primer, Portola Institute, Fricke-Parks Press, Inc., 1974
Solar Energy and Shelter Design, Bruce Anderson, Total Environmental Action, 1973
The Solar Home Book, Bruce Anderson, Cheshire Books, 1976
The Unheated Greenhouse, D. Goold-Adams, Transatlantic Arts, Inc., 1955

The Owner Builder and the Code: Politics of Building your Home, Ken Kern, Ted Kogon and Rob Thalon, Scribner, 1976

An Attached Solar Greenhouse, W.F. and Susan Yanda, The Lightning Tree Press, Santa Fe, NM, 1976

Other Homes and Garbage, Leckie, Masters, Whitehouse and Young, Sierra Club, 1975 (excellent book for explanation of heat loss calculations, properties of building materials, climate data, soil analysis)

Periodicals, Pamphlets and Papers:

A Proposed Solar Heated and Wind Powered Greenhouse and Aquaculture Complex Adapted to Northern Climates, Robert Angevine, et. al., Journal of the New Alchemists, No. 2, 1974

Optimal Shape of Greenhouse Roofs Deduced from the Solar Shape of Tree Crowns and Other Plant Surfaces, M. Bitterman and D. Dykyjova, Paper V22 from the International Congress: The Sun in the Service of Mankind: UNESCO, 1973

Greenhouse Design, Bruce Bugbee, Alternative Sources of Energy, No. 18, July 1975

Handbook of Fundamentals, American Society of Heating, Cooling and Air Conditioning (ASHRAE), 1972 (especially Chapter 59 for solar data).

High Country News, Lander, Wyoming, March 12, 1976. An article on an attached greenhouse that retains the sun's heat through energy saving devices.

Designing a Solar Greenhouse, Conrad Heeschen, The Maine Organic Farmer and Gardener, Kennebunkport, Maine, September-October 1976.

Sunworld, published quarterly by the International Solar Energy Society, P.O. Box 26, Highett, Victoria 3190 Australia

Solar Age, published monthly by SolarVision, Inc., Vernon, NJ

Solar Engineering, published monthly by Solar Engineering Publishers, Inc., Dallas, TX

Chapter IV Construction

Building Construction Illustrated, Francis Ching, Van Nos Reinhold, 1976

Your Homemade Greenhouse and How to Build It, Jack Kramer, Cornerstone, 1975

The Owner Built Home, Ken Kern, Scribner, 1975

Dwelling House Construction, Albert Dietz, The M.I.T. Press, 1971

Low-Cost, Energy Efficient Shelter, Rodale Press, 1971

How to Grow Your Own Groceries for $100 a Year, Clifford Ridley, Hawkes Publishing Company, 1974

Do-It-Yourself Housebuilding: Step by Step, MacMillan, 1973

Basic Construction Techniques for Houses and Small Buildings Simply Explained, Bureau of Naval Personnel, published by Dover

Chapter V Performance and Improvements

Books and papers listed in Appendix C and Chapter III listing.

Chapter VI The Greenhouse Garden

Winter Flowers in Greenhouse and Sunheated Pit, K.S. Taylor and E.W. Gregg, Charles Scribners' Sons, 1969

Companion Plants and How to Use Them, Helen Philbrick and Richard Gregg, Devin Press, 1966

What Every Gardener Should Know About Earthworms, published by Garden Way Research

Hydroponic Gardening, Raymond Bridwell, Woodbridge Press, 1972

Beginners Guide to Hydroponics, James S. Douglas, Drake Publishers, Inc., 1972

The Survival Greenhouse, James B. DeKorne, The Walden Foundation, 1975

How to Grow More Vegetables, John Jeavons, Ecology Action of the Midpeninsula, 1974

The Facts of Light, Ortho Book Series, 1975

The Postage Stamp Garden Book: How to Grow All the Food You Can Eat in Very Little Space, Duane Newcomb, Hawthorn, 1975

How to Have a Green Thumb Without an Aching Back, Ruth Stout, Cornerstone, 1974

The Basic Book of Organic Gardening, Robert Rodale, editor, Ballantine, 1975

Encyclopedia of Organic Gardening, Robert Rodale and staff, editors, Rodale Press, 1975 (twenty-fifth printing)

Organic Gardening Under Glass, George and Katy Abraham, Rodale Press, 1975

The Gardener's Catalogue, produced by Harry Rottenberg, William Morrow and Co., 1974

Food Gardens: Indoors, Outdoors and Under Glass, produced by Harry Rottenberg, William Morrow and Co., 1975

The Forgotten Art of Growing, Gardening and Cooking with Herbs, Richard M. Bacon, Yankee, Inc., 1972

Gardening Without Poisons, Beatrice Trum Hunter, Houghton Mifflin, 1972

A to Z of Greenhouse Plants; A Guide for Beginners, Paul Langfield, Max Parrish and Co., 1964

Gardening for People (who think they don't know how), Douglas Moon, John Muir Publications, 1975

The Gardener's Bug Book (4th Edition), Cynthia Westcott, Doubleday & Co., Inc., 1973

CHAPTER VII The State of the Art

Books and papers listed in Appendix C and Chapter III listing.

ORDER BLANK

I want to know more about Solar Greenhouses! Please send a copy of your book to:

NAME ______________________________

ADDRESS ______________________________

CITY ______________ STATE ________ ZIP ________

Send to:

John Muir Publications
P. O. Box 613
Santa Fe, New Mexico 87501

Enclosed is $6.00 plus 50 cents for postage and handling. ________
Enclosed is $6.27 plus 50 cents for postage and handling. ________
(I live in Sunny New Mexico)

P.S. Bill Yanda is available for speaking engagements. Write to us at the above address.